Language and Hemispheric Specialization in Man:
Cerebral Event-Related Potentials

Progress in Clinical Neurophysiology

Vol. 3

Series Editor: JOHN E. DESMEDT

M.D., Professor of Neurophysiology and Pathophysiology of the Nervous System,
Director of the Brain Research Unit, University of Brussels, Brussels

S. Karger · Basel · München · Paris · London · New York · Sydney

Language and Hemispheric Specialization in Man: Cerebral Event-Related Potentials

Editor: JOHN E. DESMEDT

68 figures and 22 tables, 1977

S. Karger · Basel · München · Paris · London · New York · Sydney

Progress in Clinical Neurophysiology

Cataloging in Publication
Language and hemispheric specialization in man: cerebral event-related potentials
Editor: John E. Desmedt. – Basel, New York: Karger, 1977
(Progress in clinical neurophysiology; v. 3)
1. Psycholinguistics 2. Evoked Potentials 3. Dominance, Cerebral
I. Desmedt, John E., ed. II. Title III. Series
W1 102 P963 v. 3
ISBN 3–8055–2629–6

© Copyright 1977 by S. Karger AG, 4011 Basel (Switzerland), Arnold-Böcklin-Strasse 25
Printed in Switzerland by Gasser & Cie Aktiengesellschaft, Basel
ISBN 3–8055–2629–6

Contents

Contents

Event-Related Potentials as Signs of Sensory Processing in Left and Right Hemispheres

Abbreviations

AEP	Auditory evoked potential	MAO	Monoamine oxydase
BAEP	Brain stem auditory evoked potential	MP	Motor potential
		MRF	Medial reticular formation
BP	Bereitschaftspotential	P300	Positive ERP component with peak latency of 300 msec
CNS	Central nervous system		
CNV	Contingent negative variation	REMS	Rapid eye movements sleep
EEG	Electroencephalogram	RH	Right hemisphere
EMG	Electromyogram	RL	Right-left
EOG	Electro-oculogram	RP	Readiness potential
EP	Evoked potential	RT	Reaction time
ERP	Event-related potential	R wave	Respiration-related potentials
FFP	Frequency-following potential	SEP	Somatosensory evoked potential
GSR	Galvanic skin response	SWS	Slow wave sleep
ISI	Interstimulus interval	VEP	Visual evoked potential
LH	Left hemisphere	VOT	Voice onset time
MA	Mental arithmetic	W	Wernicke area

Preface

The study of language and of higher cognitive functions of the human brain has now entered a phase of faster developments, as new and more pertinent methods of study have come into use. Data provided over about a century by the clinical observations of patients with localized cerebral lesions have raised many pertinent questions about the lateralization of functions in the cerebral hemispheres. The analysis of patients with surgical commissurotomy has more recently contributed to progress in several directions thereby placing the true dimensions of the complex problems raised into better perspective. The higher nonverbal cognitive functions of which at least a few are primarily subserved by the non-language hemisphere have only recently been given more adequate consideration.

In spite of the importance of these approaches, one must recognize that studies of the neural bases of the higher cognitive functions should be extended to intact man in whom commissural integration is possible and in whom the various parts of the two cerebral hemispheres and of subcortical structures can interact dynamically. The cerebral event-related potentials (ERPs) recorded from various points of the scalp present a distinct advantage as points of entry for such studies, within the constraints of ethical human experimentation. They appear more versatile than, say, psychological paradigms with choice reaction times.

The rationale of the ERP approach involves the recording and analysis of focal electrogeneses of various cerebral processors related to different cognitive capacities. The enthusiasm raised by such a possibility has led to a somewhat confused state because this is obviously not an area where foolproof shortcuts can be designed easily on the basis of clinical and other data. Quite a few ERP reports published over the last 8 years proposed inferential data or views which have since been challenged or disproved. In fact the ERP approach requires a strong interaction between several lines of research in order to come up with realistic and pertinent data. The identification of psychological processes involved in a cognitive capacity must be characterized to design experimental paradigms which will engage the capacity to a significant extent, more or less selectively and at a known period in time. On the other hand the analysis of ERPs from several scalp locations can only attempt identify unique focal electrogeneses which may be related to the operation of the capacity if quite a few possible artifactual contaminations of the ERPs can be excluded. Improvements in experimental designs should be facilitated by the upgrading of methodologies and of publication criteria which have been proposed after detailed discussions by *ad hoc* working parties [cf. ARDEN *et al.*, 1977; DONCHIN *et al.*, 1977a]. When studying lateralized processors it is important to ascertain which cerebral hemisphere is related to language in the subject studied, and to obtain data for control tasks which engage the right and left hemispheres to different extents.

The cerebral ERP approach must pay due consideration to all these criteria in order to achieve recognition as a unique methodology for the problems as to how activities of certain neurons and circuits get translated into processing operations which are relevant to specific cognitive behaviors. The present book has been carefully designed to attempt resolve many of the inconsistencies and past shortcomings of the ERP approach to the higher lateralized cognitive functions of the human brain. The book is the first to assemble as many as 18 chapters which present critical new data and views bearing specifically on these problems of ERP correlates of hemispheric specialization in man. A wealth of different procedures are discussed while the inadequacies of some of the earlier contributions are analysed. The book leaves the impression that this area of clinical neurophysiology is alive and well, and probably on the verge of major breakthroughs which have obvious relevance to the understanding of brain functions and to clinical applications.

J. E. DESMEDT

Ades
Kuhl & Miller
Durzwall & Whitaker
Yeni Komshian Benson
Robinson & Voreida
Webster & Webster

Begleiter & Porjesz

Keenan etal

Cited in Warren Treyler Thompson
but not in bibliography

Background and General Concepts

Language and Hemispheric Specialization in Man: Cerebral ERPs.
Prog. clin. Neurophysiol., vol. 3, Ed. J.E. DESMEDT, pp. 1–27 (Karger, Basel 1977)

The Psychobiology of Speech and Language — an Overview

ERIC WANNER, TIMOTHY J. TEYLER and RICHARD F. THOMPSON

Rockefeller University, New York, N.Y., Neurobiology Program, Northeastern Ohio Universities College of Medicine, Rootstown, Ohio, and Department of Psychobiology, University of California at Irvine, Irvine, Calif.

Speech is the one truly species-specific behavior of *Homo sapiens.* No other animal possesses language. Similarly, no other species has developed the specializations of the left cerebral cortex that make speech and language possible. It is of interest that although complex aspects of behavior such as learning and problem solving cannot be localized to any particular part of the brain, language has a relatively well-localized biological substrate. Given the localization of language function in the brain, we might expect that analysis of the neuronal mechanisms underlying language would have proceeded at a rapid pace. However, the discipline of *neurolinguistics* is only beginning. To be sure, some knowledge of the localization of speech and language function in the cerebral cortex has been available for about a century thanks to early studies by neurologists [GESCHWIND, 1970]. However, analysis of biological mechanisms underlying language requires a clear understanding of the behavior of language as studied in *psycholinguistics* which has itself existed only for some 25 years. As for aspects of brain-behavior relations, significant progress in analysis of brain substrates is dependent upon a clear understanding of the behavior.

In this introductory overview of the behavioral and neural aspects of language, WANNER summarizes the current state of psycholinguistics, with emphasis on approaches that lend themselves to analysis in biological terms. TEYLER summarizes the current state of neurolinguistics with emphasis on the evoked potential (EP) method. The fact that the brain is asymmetrical for language functions is an obvious point of departure for studies relating language processing to measures of brain activity. Recent work on aphasia has led to clear differentiations between the anterior or motor speech area of Broca and the posterior area of Wernicke. Still another research area concerns the normal development of language in the child. The invariant developmental sequence of mandatory morphemes in children was discovered only 4 years ago by BROWN [1973]. DEVILLIERS [1974] has subsequently made very good use of this critical information in a study of adult aphasics. In short, points of intersection between psycholinguistics and neurolinguistics are just beginning to develop.

I. Current Issues in Psycholinguistics

Ever since Aristotle, it has been considered fair game to draw inferences about human nature from the nature of human language. From its inception, psychology has been an active participant in this speculative tradition [BLUMENTHAL, 1970]. Occasionally, psychologists also made efforts to use empirical methods to settle questions about language and its use; but these efforts were exceptional. Only in the past 25 years has there been a sustained effort to test psychological speculation about linguistic behavior against the results of observation and experiment. During this period, a substantial psychology of language has developed which concerns itself primarily with the following questions: How is the language user able to *produce, recognize,* and *understand* the sentences of his language; and how does the child *acquire* these abilities? In the following discussion, we shall describe recent work on these issues, giving some attention to possible points of contact with neurolinguistics. First, however, a few comments about method.

Modern psycholinguistics originally exhibited an almost embarrassing diversity in its methodological assumptions, ranging from the radical behaviorism of SKINNER [1957] to the avowed mentalism of CHOMSKY [1968]. In fact, much of the early excitement generated by the field centered around essentially methodological controversies about how linguistic behavior should be studied [MILLER, 1965; KATZ, 1966]. These debates have largely subsided, leaving behind a mainstream methodology in which both behaviorist and mentalist influences are discernible. Basically, the psycholinguist sees himself as attempting to characterize the psychological processes by which linguistic knowledge is acquired and used. These processes are ordinarily conceived of as operations on information, carried out by abstract mechanisms of various design. Hypotheses about such mechanisms are tested in terms of the predictions which they support about observable behavior.

A great deal of early work in psycholinguistics [MILLER and MCNEILL, 1968] was based upon the premise that psychological models of linguistic behavior might directly employ just the statement of linguistic knowledge provided by modern linguistic theory. Roughly, the idea was that linguists would provide a description of linguistic knowledge and psycholinguists would show how that knowledge was acquired and used. However, adherence to this division of labor is now largely a thing of the past. Modern linguistics attempts to describe a language by providing an inventory of its basic units (sounds, words, word classes, phrase types, etc.) and a set of grammatical principles by which these units can be combined into meaningful sentences. Properly constructed, such principles should comprise a finite grammatical system which can provide paired descriptions of the sound and meaning of any of the indefinitely many sentences in a language. As CHOMSKY [1965, 1966] pointed out, such a grammatical system is of potential psychological interest because it characterizes what a child learns when he learns a language. However, it is important to understand that this characterization is a rather loose one. Given a limited exposure to language and a finite nervous system, the child must acquire a finite grammatical system which permits him to go beyond his own limited linguistic experience to produce and understand novel sentences in an appropriate way. But it does not follow that the child acquires just the grammatical system produced by the linguist. If the linguist's grammar is correct, then it must describe the same set of sound-meaning correspondences which the child learns.

There are, however, indefinitely many grammatical systems and there is no reason to suppose that the child's system bears a direct relation to the linguist's. On the contrary, there are several reasons to expect that a direct relationship will not occur. One runs as follows: the linguist's grammar is designed to provide the simplest possible acount of judgements which the adult language user can make about such matters as ambiguity, acceptability, and paraphrase. What the child must learn, however, is not to make linguistic judgements, but rather to acquire the ability to speak and understand sentences. It therefore seems likely that the child's grammar will be designed to accommodate these important communicative functions, which must be carried out in real time and with highly restrictive limitations upon processing resources such as memory and attention. The linguist's grammar is subject to none of these constraints. Given the gift of hindsight then, it may not be too difficult to understand the failure of experiments which tested the hypotheses that the linguist's grammar is employed directly in the process of comprehension and production [FODOR and GARRETT, 1966]. With the failure of these tests, it has become apparent that the task of psycholinguistics is not to specify how the linguist's grammar is put to use in sentence comprehension or production, but to discover what sort of grammatical principles the language user *does* employ to achieve just the degree of efficiency which he demonstrates in his linguistic behavior.

The development of the relation between linguistics and psycholinguistics may contain something of a methodological moral for neurolinguistics. Psycholinguistic data undoubtedly provide a useful characterization of the linguistic behavior which a neurolinguistic model must ultimately explain. Moreover, psycholinguistic models may provide a useful description of the functional requirements which must be met by any system capable of mediating language behavior. But it would be quite surprising if there turned out to be any simple mapping between the elements and operations of psycholinguistic models and the elements and operations of neurolinguistic models [LENNEBERG, 1973]. Clearly, neurolinguistic models must meet constraints which are not brought to bear upon the construction of psycholinguistic models. Consequently, it is often easier to see how neurolinguistic results bear upon psycholinguistic questions than vice versa.

Speech perception. When a language user perceives speech in a familiar language, he does not hear the spluttering gibberish of the acoustic signal; he hears an ordered sequence of meaningful sounds. Apparently, he knows how to translate the continuously varying acoustic signal into a series of discrete linguistic units. The major psycholinguistic questions about this translation process are these: What are the linguistic units which are recognized during speech perception? What are the acoustic cues which are used in recognition? And, what is the nature of the recognition mechanism?

Perhaps the most well-known and controversial set of answers to these questions is contained in the so-called 'motor theory' of speech perception developed at the Haskins laboratory [see LIBERMAN et al., 1967; LIBERMAN, 1970]. Motor theory begins with the assumption that the basic unit of speech recognition is the *phoneme,* a unit which linguistics defines as the minimal sound segment which a language uses to distinguish between words. From this assumption, motor theory proceeds to the more surprising claim that mechanisms of speech recognition must make use of an internal model of speech production in order to determine the way in which the acoustic representation of a given phoneme will be conditioned by its context. In this respect, the perception of speech is presumed to employ perceptual mechanisms not applied to other acoustic stimuli.

Much of the evidence for motor theory has been developed by means of the sound spectrograph [KOENIG *et al.,* 1946] and the Pattern-Playback [COOPER, 1950]. The spectrograph uses a set of electronic filters to perform a frequency analysis of the speech waveform. It produces a time-frequency-energy graph, called a *spectrogram,* which represents variation over time in the energy levels of the major frequency components of the speech signal. The Pattern-Playback is effectively a spectrograph put in reverse. Given a spectrogram as input, it synthesizes recognizable speech sounds as output.[1] By examining the spectrographic patterns in natural speech, it is possible to formulate hypotheses about the acoustic cues which distinguish speech sounds. These hypotheses can then be checked by manufacturing artificial spectrograms in which the hypothetical cue is varied while other acoustic properties are held constant. Using this technique, the HASKINS group has obtained 2 very interesting results.

The first is that speech sounds are perceived *categorically.* On most sensory dimensions, a human observer can discriminate among many more stimuli that he can identify or label absolutely [MILLER, 1956]. However, this generalization does not hold for speech sounds. In speech, discrimination is frequently no better than identification. For example, the syllables /ba/ and /pa/ differ according to whether the initial consonant is voiced, as in /ba/, or unvoiced, as in /pa/.[2] Acoustically, this distinction depends upon the size of the delay between the initial sound made by releasing the lips and subsequent sound made by vibrating the vocal folds. If this delay (which is often called *voice onset time* or VOT) is varied over a range of values in synthesized speech sounds, speakers of American English show a sharp cut-off between the 2 syllables at a VOT of about 25 msec. Sounds with shorter VOT values are uniformly heard as /ba/. Sounds with longer VOT values are heard as /pa/ [LISKER and ABRAMSON, 1970]. Moreover, although sounds which fall on different sides of this boundary are easy to discriminate, pairs of sounds which fall on the same side of the boundary are discriminated at little better than chance level.

---

[1] Speech synthesis is now generally achieved by a digital computer augmented by a device for generating waveforms from digitally specified speech spectra.

[2] The phonemic symbols used here and elsewhere throughout this chapter represent the following sounds:

/a/ *calm*	/g/ – *go*	/t/ – *to*
/ə/ – *jaw*	/i/ – *beet*	/u/ – *cool*
/b/ – *by*	/k/ – *key*	
/d/ – *do*	/p/ – *pie*	

The second discovery of the Haskins group was that speech appears to violate what can be called the *linearity* and *invariance* conditions. Since we hear a sequential procession of discrete sounds when we listen to speech, it is natural to assume that the speech signal can be divided into a sequence of temporal regions, each of which contains all and only all the cues relevant to each phoneme. However, this assumption is incorrect in several ways.

First, there are cases where the perceptual value of a given region of sound can change according to its context. For example, the stop consonants (/b/, /p/, /g/, /k/, /d/, /t/), so named because they are produced by a momentary interruption of the air flow from the lungs, typically result in a brief burst of noise when the closure is released. The frequency range of the burst provides some information about which stop consonant is being produced. However, the ranges for each stop vary according to the nature of the following vowel sound. And there is enough overlap in these ranges so that a single noise burst at 1,440 Hz is heard as /p/ before the vowel /i/ but as /k/ before /a/ [LIBERMAN *et al.*, 1952]. So, a single sound segment can represent different phonemes in different contexts.

The converse is also true: a single phoneme can be represented by different sounds in different contexts. One example of this sort is shown in figure 1, which displays spectrograms of artificial speech sounds sufficient for the perception of the syllables /di/ and /du/. The steady state bands of energy are characteristic of vowels and are called *formants*. The formant frequencies provide information about the identity of the vowels. The transitions into the formants help to signal the stop consonant preceding the vowel. Notice, however, that in this case 2 different transitions in the higher

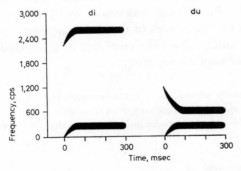

Fig. 1. Simplified spectrographic patterns sufficient to produce the syllables [di] and [du]. From LIBERMAN [1970, p. 307].

frequency formant are heard as the same phoneme /d/ in the 2 syllables. These transitions are audibly distinct. Taken out of context, the formant transition for /di/ sounds like an ascending whistle while the transition for /du/ sounds like a descending whistle [LIBERMAN *et al.*, 1967]. However, when the transitions are embedded within the context of the speech signal, they contribute to the perception of the same phoneme.

LIBERMAN *et al.* [1967] and LIBERMAN [1970] have argued that the facts about categorical perception and the context sensitivity of phonemes can be put together if we adopt an account of speech perception in which the listener uses an internal model of speech production to determine the proper phonemic analysis of the speech signal. The general claim is that the perception of a phoneme is more closely correlated with its articulation than with its acoustics. For example, in the /di/-/du/ case, 2 different sounds are perceived as a single phoneme. However, the production of both of these sounds involves a common articulation which stops the flow of air through the oral cavity by momentarily touching the tip of the tongue to the post-dental gum ridge. Therefore, the listener might achieve perceptual constancy across the /di/-/du/ syllables by determining that each of these sounds could have been produced by a common articulation. This common articulation has disparate acoustic effects because no sound is emitted at the point of closure. Only after the release is sound produced and this sound is influenced both by the location of the closure and the vowel towards which the articulatory apparatus is moving. According to LIBERMAN's proposal, the listener may use his knowledge of speech production to determine the way in which successive articulations are acoustically merged. Once recovered, these articulations presumably provide an invariant cue for the correct phonemic analysis of the signal.

The same sort of argument is applied to explain both categorical perception and cases such as the /pi/-/ka/ example described above, where a single sound can be perceived as different phonemes across contexts. Basically, the idea is that one sound can receive 2 different phonemic analyses if it can be produced by 2 different articulations. Similarly, categorical perception occurs where local regions of a stimulus continuum are produced by distinct articulations. Since by hypothesis, perception is determined by articulation, it follows that these regions will also be perceived distinctly.

The motor theory of speech perception has received several objections. First, it is questionable whether there are in fact articulatory invariants associated with phonemes [FODOR *et al.*, 1974]. Electromyographic recordings of muscle potentials in the speech organs have generally failed to show invariant patterns associated with each phoneme [MCNEILAGE, 1970]. Thus, it may be that the articulation of a given phoneme is also conditioned by context. Second, there is now some evidence that categorical perception can be explained strictly in terms of acoustic cues without reference to articulation. Thus, STREETER [1974] has pointed out that the VOT boundary between /pa/ and /ba/ contains a number of acoustic properties which might provide the basis of a discrimination. Moreover, ADES [1974] and MILLER *et al.* [1974] have shown that categorical perception can be obtained at approximately the same VOT boundary found in English with synthetic stimuli which are not recognized as speech sounds.

Finally, although the Haskins' results show that the conditions of linearity and invariance are violated by the acoustic representation of phonemes, their results do not show that these conditions are also violated at the level of the syllable. So, for example, the 2 distinct sounds specified in figure 1 convey the same initial phoneme; however, they convey different syllables. Therefore, if the listener recognizes speech syllable by syllable, the context sensitivity problem posed by the /di/-/du/ case disappears. Similar arguments can be constructed for the other well known cases of phonemic context sensitivity. Moreover, the syllable can be shown to be generally less sensitive to context than the phoneme. For example, although it is impossible to synthesize intelligible messages by splicing together phoneme-sized segments of natural speech [HARRIS, 1953], more acceptable results can be achieved with segments which approach the size of the syllable [PETERSON *et al.*, 1958]. Therefore, if the syllable, rather than the phoneme, is the basic perceptual unit, speech perception might not require reference to an internal model of production in order to disentangle the effects of context. RUBIN [1974] has recently summarized several lines of behavioral evidence which demonstrate that the syllable hypothesis is psychologically tenable.

Neurolinguistic evidence has played a larger role in the controversy over the motor theory of speech perception than it has in most psycholinguistic issues. Besides the EMG research cited above, both clinical evidence concerning linguistic disorders and experimental work with the electroencephalographic recording techniques have been involved. For example, LENNEBERG's [1962] case of *congenital anarthria* is sometimes cited as providing difficulties for motor theory [FODOR *et al.*, 1974]. That patient could not coordinate the organs of speech and was unable to produce any but the most infantile vocalizations. However, his ability to perceive and understand speech was unimpaired. He could respond appropriately to complex demands, even when these were delivered over a tape recorder. This observation appears to constrain motor theory in the following way: Motor theory claims that speech is perceived by correlating acoustic input with the motor commands necessary to reproduce that input. LENNEBERG's patient cannot have learned this correlation by monitoring his own speech. Therefore, if motor theory is correct, he must have learned the correlation in some other way (which is difficult to specify), or the correlation must be innate.

A second example of relevant neurolinguistic work falls on the other side of the issue. Dichotic listening experiments generally show that speech stimuli are perceived more accurately in the right ear than in the left [STUDDERT-KENNEDY and SHANKWEILER, 1970]. Non-speech stimuli do not show the same asymmetry. WOOD et al. [1971] have shown that these behavioral results are congruent with the auditory EP obtained during speech perception. WOOD [this volume] points out that the EP results are consistent with the idea that speech perception requires a specialized linguistic decoder and that this decoder is localized in the left hemisphere. Note, however, that nothing about this result requires the stronger conclusion that such a linguistic decoder should make use of the neurological mechanism involved in speech production.

Comprehension. It is a common impression that the meaning of a message is easier to retain than its exact wording. This impression has been confirmed by a variety of psycholinguistic experiments [FILLENBAUM, 1971]. Moreover, it has been shown that unless subjects make an explicit effort to memorize the particulars of a message, information about its exact wording is lost almost immediately after comprehension [WANNER, 1974]. Thus, it appears that the process of comprehension results in an abstract representation of meaning and that only this abstraction is ordinarily consolidated in long-term memory.

Psycholinguistic research on comprehension divides roughly into 2 types. First, there has been an attempt to determine the nature of the stored semantic result of comprehension, thereby providing a psychological characterization of meaning. And second, there has been an effort to build and test models of the comprehension process.

Research of the first type consists largely of memory experiments aimed at determining exactly what kind of meaningful information is extracted from a verbal message and retained in memory. These results have confirmed a number of interesting characterizations of semantic memory. Among the most important are the following:

A. Semantic memory is relational: It retains information of the type which specifies 'who did what to whom', although it typically does not retain any information about the particular arrangements of words which convey such relationships. For example, if subjects in a memory test are given a sentence such as,

1) John sold the book to Mary,

they will typically recognize the difference between (1) and any subsequently presented sentence which conveys a different set of semantic relations, such as:

2) Mary sold the book to John.

However, they will ordinarily fail to recognize the difference between (1) and any of the following sentences, all of which preserve semantic relations:

3) The book was sold to Mary by John.
4) John sold Mary the book.
5) Mary bought the book from John.
6) The book was bought from John by Mary.

B. Semantic memory is integrated: The semantic information from a given sentence may be combined together with information previously existing in memory to provide a coherent representation of an entire scene or episode. For example, BRANSFORD and FRANKS [1971] have shown that the information conveyed in several simple sentences describing different aspects of a single episode is apparently integrated in memory with the result that a subject who has been exposed to the set of simple sentences will often mistakenly recognize a single complex sentence which describes the whole episode.

C. Semantic memory is inferential: In addition to containing the relational information explicitly conveyed by the linguistic message, semantic memory may also contain information about relations which can be logically inferred from the message. For example, BRANSFORD et al. [1972] found that if subjects exposed to a sentence such as,

7) Three turtles rested on a floating log and a fish swam beneath it,

they will be unable to recognize the difference between (7) and a new sentence such as:

8) Three turtles rested on a floating log and a fish swam beneath them.

Presumably, the confusion between these sentences is due to the fact that (8) is inferable from (7). When this inference is removed by changing the preposition 'on' to 'beside' in both sentences, the confusion disappears. By themselves, these results do not require the conclusion that inferences are made during comprehension and then stored. It might be the case that the inference which leads to the confusion between (7) and (8) is made only during retrieval. However, CHARNIAK's [1972] analysis of the comprehension process strongly suggests that at least some inferential work must be performed during comprehension. Experimental research aimed at determining the nature and extent of such inferences has begun [KEENAN *et al.*, unpubl.].

The second type of psycholinguistic research concerning comprehension investigates the psychological processes involved in determining semantic relations, integrating information over sentences, making inferences, and so forth. By far the most developed line of research of this sort concerns the recovery of semantic relations. Much of the interest in this problem stems from CHOMSKY's development of transformational grammar [CHOMSKY, 1957, 1965] which indicated that the problem of determining semantic relations was non-trivial, but that an elegant solution was possible. A rough outline of his argument runs as follows:

CHOMSKY assumes that there are a number of grammatical functions in any sentence which provide essential information about the semantic relations it conveys. For example, in a simple sentence with a verb of action, the noun phrase which functions as the *subject* of the sentence names the *agent* of the action. CHOMSKY points out, however, that the cues which supply evidence about the grammatical function of a phrase are exceedingly complicated. There is no simple one to one relation between the structural location of a phrase and its grammatical function. For example, in a simple active sentence such as,

9) The policeman found the package,

the subject is in initial position. But in a passive version of the same sentence, the subject will appear after the verb and within a prepositional phrase, as in:

10) The package was found by the policeman.

So, a single grammatical function can be fulfilled by phrases appearing in 2 different structural positions in different contexts. The reverse is also possible: a single structural position may be related to several different grammatical functions. As evidence, notice that sentence 10 is ambiguous;

its meaning depends upon whether 'the policeman' functions as subject, or as a locative adverbial indicating where the package was found. These examples suggest that no uniform structural definition of the grammatical functions is possible. If so, the listener cannot determine a grammatical function of a phrase on the basis of any invariant structural property.

However, in a transformational grammar, the feasibility of a uniform structural definition is formally preserved. The preservation is accomplished by introducing an abstract level of syntactic description, called *deep structure*, which represents the grammatical functions of a sentence in an invariant configuration. Transformational rules are then applied to the deep structure of a sentence in order to obtain a description of the observable arrangement of the phrases in the sentence, sometimes called its *surface structure*. In this way, transformational grammar offers an elegant solution to the problem of defining grammatical functions, and this solution was initially carried over as a psychological hypothesis about comprehension. Crudely put, the idea was that since grammatical functions can be presented at the level of deep structure and, since deep structure is related to surface structure by transformational rules, then (by hypothesis) the listener might determine grammatical functions during comprehension by means of some sort of internal application of the transformational rules.

In the middle 1960s, this hypothesis was tested in a series of experiments designed to determine whether the number of grammatical transformations required to relate the deep and surface structure of a sentence is correlated with general degree of difficulty which the listener has in understanding that sentence [FODOR *et al.,* 1974]. This prediction was tested in several different experimental paradigms involving measures of long-term memory [MEHLER, 1963], short-term memory [SAVIN and PERCHONOCK, 1965], as well as measures of the reaction times required to perform such tasks as evaluating the truth of falsity of a given sentence [MCMAHON, 1963; GOUGH, 1965] or converting a sentence from one syntactic form to another [MILLER and MCKEAN, 1964]. These early studies generally tended to support the transformational hypothesis: sentences with a greater number of transformations in their derivational histories tended to be harder to recall and to take more time to respond to. Subsequent reviews of this work, however, have demonstrated that it is at best inconclusive. In general, it has been shown that these tests of the transformational hypothesis failed to control for variables correlated with transformational complexity, in particular sentence length [MATTHEWS, 1968] and meaning [FOA and SCHLESINGER, 1965]. Where such variables have been controlled, evidence for the transformational hypothesis tends to evaporate [FODOR and GARRETT, 1966].

The demise of the transformational hypothesis has generally led psychologists to abandon the attempt to preserve a uniform structural definition of grammatical functions. In place of the transformational model, several schemes have been proposed which determine grammatical functions directly from surface structure. Since grammatical functions are displayed in different arrangements in different surface structure contexts, these schemes necessarily apply context-sensitive definitions of grammatical functions.

Perhaps the most promising and complete scheme of this sort is the so-called *augmented transition network* (or ATN), originally designed by THORNE *et al.* [1968] and since developed by BOBROW and FRAZER [1969], WOODS [1970], and KAPLAN [1973]. The ATN is basically a notation for building computational systems which are capable of determining the grammatical functions of the phrases in a sentence. Such systems can be set up as a computer program, the performance of which can be compared against human performance in comprehension. The generality of ATN notation permits the psycholinguist to identify and manipulate parameters of the system in order to fit the facts of human performance.

For example, there is now a good deal of psycholinguistic evidence which shows that comprehension is essentially a sequential process in which information at the beginning of a sentence is handled before information which arrives subsequently [HOLMES and FORSTER, 1970; MARKS, 1967]. When an ATN is configured to operate sequentially, it works its way through an input sentence word by word formulating hypotheses about the contextually appropriate syntactic category of each word, the proper grouping of words into phrases and the grammatical function of each phrase. Since these hypotheses are formulated on the basis of surface structure cues, their success is necessarily dependent upon context. If an ATN operates sequentially, computational problems arise whenever the context which determines the fate of a given hypothesis follows the word or phrase which the hypothesis concerns. In such cases, the ATN must have the ability either to formulate an initial hypothesis and revoke it later if subsequent context requires, or to suspend judgement and await the arrival of the relevant evidence.

It is not difficult to show that the application of either of these strategies places an additional processing load on the ATN. In a recent series of experiments, WANNER and MARATSOS [1974] have attempted to determine whether the problem of subsequent context also places an additional processing load on the human listener. These experiments employed sentences with relative clauses containing a critical region in which the subsequent context problem is particularly acute for an ATN. WANNER and MARATSOS measured processing load within such sentences by inserting an extraneous task at various points within the presentation of the sentence, and observing the amount of mutual interference between performance on the task and the comprehension of the sentence. They found that interference increased sharply within the critical region of the relative clause, thus suggesting that the subsequent context problem increases processing load for the human listener as well as for an ATN. Subsequent experiments in this series have been designed to see whether the listener handles the subsequent context problem by suspending judgement or by formulating tentative hypotheses and correcting them later if necessary [WANNER and SHINER, unpubl.].

To date, the psycholinguistic study of comprehension has made almost no contact with neurolinguistics. However, it is not difficult to see how neurolinguistic techniques might be usefully applied to this study. For instance, early psycholinguistic hypotheses about comprehension mechanisms were tested almost exclusively in terms of general differences in the difficulty of processing different sentence types. As psycholinguistic models have become more sophisticated, it was possible to formulate hypotheses about variables, such as processing load which fluctuate continuously over the course of a sentence. However, these predictions can be difficult to test behaviorally. For example, if processing load is measured by observing interference from a concurrent task, it is usually necessary to present a given sentence (or different instances of a given sentence type) many times, varying the location of the extraneous task across presentations, in order to obtain a profile of processing load in a single sentence (or sentence type). Multiple presentations are not only laborious; they may also induce subjects to develop special purpose strategies which give misleading results. Such strategies may also develop in response to the extremely heavy and unusual demand characteristics of experiments which employ concurrent tasks. In general, there can be no assurance that the human listener ordinarily processes sentences in the way a subject does in an experiment in which he must both process a sentence and do something else at the same time.

The problems which attend behavioral measures of processing load might be eliminated if physiological indicants of processing load could be developed. Physiological measures tend to be less obtrusive than behavioral measures and they provide continuous output. It remains to be seen, however, whether useful physiological measures can be achieved. Several efforts to measure physiological changes during sentence comprehension have been made. The variables measured include pupil dilation [WRIGHT and KAHNEMAN, 1971; STANNERS et al., 1972], GSR [BEVER et al., 1969] and scalp-recorded EP [this volume]. In several of these experiments, some physiological sensitivity to linguistic differences between sentence types has been demonstrated. What is still unclear, however, is whether physiological measures will prove responsive to subtle changes in cognitive states which may occur at a rapid rate during sentence comprehension.

Language acquisition. Most children utter their first intelligible words before their first birthday and begin to show rudimentary linguistic comprehension at about the same time. During their second year, they ordinarily begin to combine words together to form interpretable expressions of 2 or 3 words in length [BROWN, 1973]. By the age of 4, according to some authorities [MCNEILL, 1968], the acquisition of syntax is largely complete. The development of semantics obviously takes a great deal longer.

This rapid and regular acquisition of the complexities of human language is probably the most interesting of psycholinguistic phenomena. It is also the most mysterious. There are many questions, but few (if any) conclusive answers. Here is a sampling:

A. What is the function of experience in the process of language acquisition? Children who are totally deprived of linguistic exposure do not spontaneously develop language [c.f. BROWN, 1958; FROMKIN et al., 1974]. Institutionalized children, whose commerce with language is greatly limited, show retarded linguistic development, particularly in the early years [LENNEBERG, 1967]. Exposure to language is obviously a necessary condition for acquisition. However, it is not at all clear how experience plays its necessary role. Attempts to accelerate development by means of explicit training have yielded only temporary effects [LURIA and YUDOVICH, 1959] or no effects at all [CAZDEN, 1965]. In fact, CAZDEN found that daily periods of free conversation with an adult stimulated faster grammatical development than training sessions in which the child received an immediate and fully developed adult paraphrase of each of his fragmentary utterances.

Results such as these are sometimes cited as evidence that the primary function of experience is not to reinforce linguistic behavior, but simply to provide linguistic examples which the child may use to test his hypotheses about the structure of his language [FODOR et al., 1974]. This view may ultimately prove to be correct; however, the possible role of reinforcement probably should not be so readily dismissed, particularly if we broaden the notion of reinforcement to include any type of feedback from parental responses. It is true that parents rarely praise or correct their children in response to the grammatical well-formedness of their children's utterances [BROWN, 1970]. However, they respond to these utterances in many other ways. It seems likely that such responses will become more and more congruent with the child's communicative intentions as the child's utterances become more well formed and therefore more interpretable to the adult. Thus, parental responses which do not explicitly praise or punish may still convey linguistic information. At present, however, it is quite unknown whether this type of parental feedback is either a necessary or important dimension of experience.

B. How is the acquisition of language constrained by innate factors? Obviously, it would be impossible to acquire a language based upon speech sounds which are beyond our constitutional abilities to hear or produce, nor could we acquire a language which conveys conceptual distinctions that we are unable to appreciate. Innate factors obviously do constrain the nature of the language that we acquire and use. The question, however, is whether language acquisition is shaped by genetic factors which are unique to language. Is there an innate, language-specific acquisition mechanism, or is language acquisition constrained only by general physical and mental abilities? There are several sorts of evidence which bear upon this question. All are indirect and quite indefinite. Here are some examples:

LENNEBERG has argued that comparisons of the human speech organs with the homologous structures in the higher primates reveal differences which suggest that 'some aspect of sound-making efficiency might... have played into the mechanisms of natural selection during the history of the (human) species' [LENNEBERG, 1967, p. 71]. The problems of neuroanatomical evidence of linguistic specialization is considered elsewhere in this volume.

KATZ [1966] and others have argued that the existence of common linguistic features in historically unrelated languages [the so-called linguistic universals; GREENBERG, 1963] constitutes evidence for an innate language-specific acquisition mechanism which select for each feature. However, the question arises whether there are no other cognitive or environmental factors which might account for the observed convergence among languages. Such a demonstration is clearly difficult, if not impossible. Moreover, in some cases, a plausible account of linguistic universals can be obtained from specifically non-linguistic cognitive factors, such as limitations on short-term memory.

EIMAS *et al.* [1971] have recently shown evidence of what appears to be categorical perception of speech sounds in 4-week-old infants. EIMAS' technique takes advantage of the fact that infants can be trained to make a sucking response in order to obtain the sensory stimulation provided by a picture or a sound. If training continues without change in the stimulus, the frequency of sucking will gradually decrease over trials. If a new stimulus is employed, the frequency of sucking will return to initial levels. Thus, changes in sucking frequency can be used as an indicant of the infant's perceptual discriminations. EIMAS used this technique to discover how infants discriminated among artificial stop consonants which varied in equal steps along the VOT dimension (see above). Quite surprisingly, EIMAS found that infant discriminations match those of adults. If a pair of sounds fell into distinct adult categories, infants gave evidence of discriminating between them. However, infants would not distinguish a pair of sounds which fell in the same adult category. This result has sometimes been taken to mean that the preverbal infant categorizes speech sounds and that his categories match adult categories [FODOR *et al.,* 1974]. If true, this result would provide strong evidence for an innate, language-specific perceptual system. However, this interpretation of EIMAS' result may be incorrect. As noted earlier, the adult category boundaries appear to lie at points along the VOT dimension which are of high acoustical prominence. Therefore, the infant may simply be responding to an acoustic distinction. Phonemic categories may have gravitated toward these boundaries precisely because of their acoustic distinctiveness. If this account is correct, then EIMAS' results do not provide evidence for an innate, language-specific perceptual mechanism.

This notion has received support from the demonstration by KUHL and MILLER [1975] that the chinchilla displays categorical perception. The experiment was conceptually similar to that reported by EIMAS. The chinchilla, a mammal with auditory capabilities close to man's, discriminated between synthetic human speech differing in the VOT dimension. This observation lends further support to the idea that phonemic categories have been selected to be highly distinctive to the auditory apparatus and are thus easily perceived by the chinchilla.

C. Is the ability to learn language species-specific? Studies on chimpanzees: Until recently, this question could always be given a quick affirmative answer. Animal communication systems appear to lack important design features of human language [McNEILL, 1970] and all efforts to teach human language to non-human species had ended in failure [KELLOGG, 1968].

Recently, however, there have been 3 separate reports of success in teaching chimpanzees to master communication systems which bear some resemblance to human language [GARDNER and GARDNER, 1971; PREMACK, 1971; RUMBAUGH et al., 1974]. In each case, success was achieved by replacing the vocal-auditory communication channel with a system for producing and receiving linguistic signs which is more congenial to chimpanzees. Although the 3 methods differ greatly both in training techniques and outcomes, the general results can be crudely summarized in a single list.

First, it appears that chimpanzees can learn arbitrary signs for classes of objects, actions, and relations. As many as 85 signs have been mastered by one chimp. Second, chimpanzees will spontaneously put signs together in short sequences which can be sensibly interpreted in terms of semantic relations found in human language. Third, chimps can be trained to use longer sequences of signs in a limited vocabulary to represent propositions, questions, requests, and implications.

Impressive as these accomplishments are, there are a number of reasons to question the notion that these chimps have mastered language, or even its rudiments [BROWN, 1973; FODOR et al., 1974]. First, it is not clear that chimpanzees can spontaneously use syntactic cues, such as sequential order, to convey semantic relations. Where chimpanzees combine signs spontaneously, the combinations appear to be order free. This leaves some doubt as to whether the combination of signs represents an underlying combination of concepts or whether it is an accidental concatenation. Where chimpanzees have been rigorously trained to combine signs in a fixed order, they do not use these combinations in spontaneous communication.

Second, the communication systems mastered by these chimpanzees lack the productivity of human language in several respects. They have no phrase structure: a single term cannot be replaced by a phrase in order to provide a more elaborate characterization of some item in an expression. They lack recursion: one expression cannot function as part of another expression. They also lack what BROWN [1973] has called systematic scope. A chimpanzee may master a certain grammatical principle, but its application will be restricted to a narrow range of forms. In human language, many rules apply across the board to all forms in the system.

To summarize then, the question of whether language is species-specific still deserves an affirmative answer. However, the 'linguistic apes' may have made us somewhat more sophisticated about just what aspects of language are reserved to our species.

II. Neural Aspects of Language Behavior

Neurophysiology and physiological psychology have learned a great deal about the operation of sensory and motor systems, but the knowledge of the integrated functions of the non-primary and association areas of the brain is quite rudimentary. We know something about the role of certain association areas in governing behavior. We cannot yet reconstruct 'recepto-topic' maps of these areas, nor can we in most cases describe the necessary and sufficient stimuli to activate many of the neurons in these areas. As one ascends the phylogenetic scale, the amount of neocortex devoted to associa-tion areas increases dramatically. One of the most complex behaviors exhibited by man is linguistic behavior. Of all the species in which commu-nication has been studied, none even begins to approach the degree of complexity and ubiquity of the human species. Similarly the amount of association cortex is nowhere as fully developed as in man. It has been demonstrated that a significant portion of the association areas of man is implicated in language behavior. Some of the evidence will be summarized below. The reader is also referred to a review of neurolinguistics by DING-WALL and WHITAKER [1974]. These authors have carefully reviewed a much wider range of topics than are treated in this chapter. Their coverage includes neurosurgery, electrical brain stimulation, electromyography and behavioral studies.

Historical aphasiology. Neurophysiologists and physiological psycho-logists have been reluctant to approach the association cortex as a neural substrate for the analysis of behavior because of the sheer complexity of the tissue and lack of basic understanding of its organization and function. However, sufficient progress has been made in allied fields to allow investiga-tions to proceed on the association areas of the human neocortex with some promise of elucidating their function [SCHUELL *et al.,* 1964; GOODGLASS and BLUMSTEIN, 1973]. Most of the information regarding the function of association cortex in man has come from clinical studies of pathology or lesion. The underlying assumption is that there are areas of cortex in which it is possible to localize language function. In 1861, BROCA discovered that lesions restricted to the posterior portion of the frontal pole of the brain resulted in deficits in the ability to use words. The general term given to these and other disorders of language is 'aphasia'. Aphasia refers not to damage to the muscles of articulation, but rather to more central deficits occurring within the confines of the brain. In addition, aphasic defects are not necessarily accompanied by problems in any other modality or intellectual capacity. Broca's aphasia is characterized by slow and laborous speech. The speaker has obvious difficulty in articulation and correct pronunciation. Patients often omit words and endings of words, resulting in a staccato-like verbalization of material. However, understanding of spoken or written material is normal.

The Broca's aphasia patient exhibits quite adequate musical capacities even to the point of correctly singing a melody with some elegance, while totally failing to produce a word or sentences with clarity. These observation have led to the suggestion that Broca's area is concerned with the phonological and the syntactic aspects of language generation. Thus, the impetus for the possibility of studying the cerebral localization of specific language functions, relatively untainted by other considerations, was given some credence by these early studies. Since the time of BROCA, as GESCHWIND [1970] pointed out, it has been determined that it is not always possible to delimit the area of cortex resulting in the symptoms currently known as Broca's aphasia. Nonetheless, BROCA observed that lesions on the non-dominant hemisphere had not such detrimental effects on language. In 1874, WERNICKE associated lesions generally limited to the posterior and lateral portions of the temporal lobe of the dominant hemisphere with a failure to understand language and a deficit in language production. The individual with lesions in Wernicke's area can produce quite lengthy sentences that are totally devoid of meaning. The sentences are syntactically correct and properly articulated, but are semantically empty. The patient often speaks rapidly, so that an individual not listening closely to the content of the speech may judge it to be entirely normal. However, the content is entirely inappropriate. He often substitutes inappropriate words or closely associated words for the word which he is unable to produce. Whereas lesions in Broca's area are not accompanied by deficits in understanding of spoken or written language, the patient with Wernicke aphasia suffers from severe loss of understanding of verbal materials, although the perception of musical sounds and other non-vocal auditory signals is apparently normal [GESCHWIND, 1970; GOODGLASS, 1972]. Table I presents clinical examples of the language produced by damage to Broca's area and Wernicke's area. The latter, which is concerned with the comprehension of language, has been viewed as connected anatomically with Broca's area which is primarily concerned with the production of language, and WERNICKE [1874] predicted, on the basis of such views, that interruptions of these connections would produce particular clinical syndromes. For example, in word deafness the patient, who is normal by audiometric measurements, completely fails to

Table I. Transcripts of the speech of persons suffering from (A) damage to Broca's area, and (B) damage to Wernicke's area

---

A. I am taking – ah – Sherring's mixture (laugh). It's easier for me to – stalk – talk staccato and breaking up each word – into sentences – breaking up sentences into words provided they have – ah – not many syllables. Ah, syllable is hard – Precise words that I have trouble with are Republican and Epics – Epis – capalian. From LENNEBERG [1967, 195].

B. Well, I thought thing I am going to tell is about my operation and it is not about all I can tell is about the preparation the had was already the time was when they had me to get ready that is they shaved off all my hair was and a few odd parts of pencil they give me in the fanny. From PENFIELD and ROBERTS [1959].

---

comprehend spoken language. He has the capacity to produce normal language and also to comprehend written language, but fails totally with spoken material. Examinations of the area of damage to the brains of these patients reveals that the fiber pathways from the auditory areas of the cortex into Wernicke's areas are totally destroyed on both sides of the brain, isolating Wernicke's area from the auditory input. Other syndromes have been described by GESCHWIND [1970] and others, attesting to the general utility of this neuroanatomical model of language production in the human (disconnection syndromes).

Contemporary studies of the cerebral basis of language behavior. Analysis of individuals suffering from various types of brain damage by contemporary workers have provided a good deal of information regarding the neural basis of language [GESCHWIND, 1970; LURIA, 1974; SCHUELL *et al.,* 1964]. Recent advances have been made by studying individuals suffering from surgical or accidental lesions of the corpus callosum and other connections between the hemispheres. By observing the type of deficit and relating it to the known or clinically determined site of lesion, knowledge has been gained regarding the hemispheric localization of language functions [GAZZANIGA *et al.,* 1965]. On the anatomical front, evidence has been obtained showing a hemispheric asymmetry in the absolute volume of the cortex in the vicinity of the traditional language areas. For example, the area lying in a region corresponding to that of Wernicke's area is of greater size on the dominant side of the human brain than on the non-dominant side [GESCHWIND and LEVITSKY, 1968; DAVIS and WADA, this volume]. However such asymmetries, in themselves, are not necessarily indicative of linguistic asymmetry. LeMAY and GESHWIND [1975] found anatomical cerebral asymmetries in the brains of the great apes (particularly the orangutan) similar to those seen in man. Similar anatomical asymmetries are seen in the chimpanzee brain [YENI-KOMSHIAN and BENSON, 1976]. The functional significance of these findings are unknown but may be related to the anatomical asymmetry seen in the visual cortex of the cat [WEBSTER and WEBSTER, 1975] which has been related to a functional behavioral asymmetry by ROBINSON and VONEIDA [1975].

Recent examinations of the quantitative aspects of language deficits in patients suffering from Broca's aphasia by DEVILLIERS [1974] show the degree to which neuroanatomical and neurophysiological insights can be gained from purely behavioral data. As noted earlier, a feature of Broca's aphasia is the omission of certain words (function words such as articles and prepositions) and word endings. BROWN et al. [1964, 1973] have found that children also omit these function words during certain stages in language acquisition. BROWN noted that the order of appearance of the correct use of 14 of these grammatical morphemes was consistent across children. This observation of a developmental invariance in function word acquisition could reflect brain maturational sequences underlying this behavior. DEVILLIERS analyzed the difficulty ordering of these same 14 grammatical morphemes in adult patients suffering from Broca's aphasia. She found a stable ordering of difficulty across patients which differed from the invariant ordering of acquisition as seen in children.

The observation of a low correlation between ordering of difficulty and ordering of acquisition has implications for the neurological substrates of this behavior. It is safe to assume that the area(s) of injury to each of the aphasics' brains are not identical. Nevertheless, the ordering of difficulty is similar among patients. If restricted portions of the brain were primarily responsible for each of the grammatical morphemes, this result would not be predicted. In this light, we tentatively suggest that at least these grammatical morphemes are not represented in terms of discrete cerebral loci. If the process of language acquisition in the child builds upon previously established grammatical morpheme cell assemblies, it might be expected that those appearing last would involve the most intricate neuronal organization and would be the most susceptible to interruption by a functional disorder. This, however, does not appear to be the case. The aphasic patient does not experience the most difficulties with those grammatical morphemes acquired late in the developmental sequence. One must interpret these differences with caution, as DEVILLIERS points out, because on the one hand the data reflect behavioral and cerebral immaturity, while on the other hand they are indications of functional disorder. There is no single variable, including developmental acquisition ordering, that satisfactorily accounts for these differences. However, theories of the cerebral substrates of language must be able to deal with these findings. In any case, it must be pointed out that studies of brain lesions or tumors suffer limitations from the difficulties of determining the actual extent of damage. Incomplete post-mortem studies on the brain, presence of diffuse changes (compression by tumor, lesions during surgery, etc.) and other difficulties inherent to work on man are to be taken into account when discussing these data. It should be remembered that a number of scholars emphasized the difficulties of attempting to localize language functions [HEAD, 1926; LURIA, 1974; LENNEBERG, 1973].

Electrocortical indices of hemispheric dominance and language. EP studies have recorded electrical activity from the scalp overlying language areas and showed evidence of EP differences in the dominant hemisphere when comparing linguistic and non-linguistic stimuli [WOOD *et al.,* 1971; MORRELL and SALAMY, 1971; McADAM and WHITAKER, 1971; GAZZANIGA and HILLYARD, 1971; BUCHSBAUM and FEDIO, 1970; NEVILLE, 1974; see this volume]. Studies of the spontaneous electroencephalogram (EEG) showed less alpha rhythm generation in the dominant cortex during verbal tasks as compared to spatial tasks, while non-dominant hemisphere alpha suppression was recorded rather in the spatial tasks [GALIN and ORNSTEIN, 1972; ROBINS and McADAM, 1974; McKEE *et al.,* 1973; see this volume].

In summary, the results of the above experiments indicate that electrophysiological measures of cortical processes associated with linguistic behavior show a predominance of activity in the dominant hemisphere and a further specialization of activity as seen in certain regions of the dominant hemisphere of the brain.

Electrocortical correlates of semantics. Studies on semantic correlates of linguistic behavior as reflected by electrophysiological measures of brain activity are also available. MATSUMIYA *et al.* [1972] studied auditory evoked potentials (AEP) and noted that meaningful noise stimuli showed a larger AEP amplitude difference between the 2 hemispheres than did relatively meaningless words to which the subject was not required to respond. This points to the importance of the semantic content of the auditory stimulus as a determinant in hemispheric AEP differentiation. It has been of course known for some time that non-linguistic auditory stimuli can be cognitively important and can elicit so-called 'decision' components in the cerebral EPs [cf. DONCHIN and LINDSLEY, 1969; DESMEDT, 1977b]. TEYLER *et al.* [1973] reported hemispheric asymmetries in AEP to verb and noun meanings of ambiguous words such as 'fire' (which can be a verb or a noun). The AEP were actually evoked by a probe stimulus (click) which followed the word. The AEP wave forms were different in any hemisphere for verb or noun forms of such ambiguous words, but similar for any of them between the 2 hemispheres (although larger in the dominant hemisphere). The assignment of meaning to an ambiguous word may have an influence on the AEP wave form. BROWN et al. [1973], utilizing a similar ambiguous word paradigm, demonstrated that analysis based upon the verbally presented stimulus word itself rather than a click following a word resulted in AEP correlates of meaning similar to those found by TEYLER et al. [1973] – see also ROEMER and TEYLER [this volume].

Following upon these first demonstrations of electrocortical correlates of semantic processing have come several interesting demonstrations of the involvement of language areas of the dominant hemisphere in other semantic paradigms. Utilizing an ambiguous stimulus paradigm, BEGLEITER and PORJESZ [1975] demonstrated evoked potential differences related to cognitive decisions about the physical attributes of the stimulus. This paper provoked some comment regarding both methodology and interpretation [DONCHIN, 1975]. BROWN et al. [1976], utilizing the ambiguous phrase, found maximal evoked potential differentiation over the anterior speech area, a surprizing result that may indicate a larger role in speech processing than previously assumed.

Some of the most fascinating and imaginative experiments are reported by CHAPMAN [1973, and this volume]. He has employed word items from the Semantic Differential of CHARLES OSGOOD as his stimuli. The Semantic Differential is based upon a large body of normative data on word meaning across languages. Through factor analysis OSGOOD found that three dimensions of connotative meaning were sufficient to account for a good portion of the total variance in judged meaning. CHAPMAN employed words which were neutral on two scales and highly polar (+ or −) on the third scale (the E or evaluation scale). These word stimuli were then presented to subjects in random order. The averaged electrocortical response to E+ and E− words was different – a remarkable finding considering that each stimulus word was different and only presented once. According to CHAPMAN the only difference between E+ and E− words was in their connotative meaning. All of the above experiments have dealt with language understanding. Characteristic electrocortical activity has also been shown to correlate with overt and covert speech as has been demonstrated by PINNEO [1976].

Limitations to neurolinguistic research. Although neurolinguistics is in its initial stages, several limitations to the study of brain correlates of language have already emerged. The problems are of 3 sorts. First, the EP wave forms obtained from the scalp in subjects exposed to, or producing, complex linguistic information are not readily amenable to quantification. The second limitation deals with the current necessity of averaging AEPs to the same linguistic stimuli. The third constraint is a task related consideration due to the nature of the response required on the part of the subject. For example, many of the linguistic stimuli used to date have a physical reference in the environment which can be mentally imaged. This leads to the suggestion that many of the attempts at dealing with brain correlates of language are in fact dealing with processes such as visual imagery rather than linguistics.

A fundamental problem facing investigations of AEPs to linguistic or other complex stimuli is the absence of any suitable metric to express the complex waveform and to assess the differences between waveforms. The metrics devised and employed to date include discrete amplitude measures, power spectra, correlations, linearity measures, and repeated point analyses of variance. While some of these metrics are quite sophisticated, they all are less than ideal in that sizeable portions of the information contained within the waveform are disregarded. Procedures should be developed to adequately quantize the complex waveforms obtained. A suitable metric must take into consideration the following: the power spectra of the EPs, the phase relationship between waveforms, the amplitude of the components of the waveform, and the amount of variance exhibited by the components of the waveform. Such a metric should enable one to quantify and precisely characterize the information represented by the complex potential recorded from the surface of the brain.

The second fundamental obstacle to research in the area of neurolinguistics lies in the current requirement to deal with average EPs to linguistic stimuli, thus implying the repeated presentation of the same stimulus. Repeated presentations of linguistic stimuli are certainly not normal occurrences in natural language. Even utilizing procedures designed to minimize the effects of repetition, such as employing distractor words and embedding the target word within a matrix of non-target words, subjects find it very difficult to consider the stimulus word in the same manner across repetitions. For example, consider repeated visual or auditory presentations of a stimulus word such as 'cat' in which the subject is instructed that the word refers to a domestic house cat. Individuals find it very difficult to focus on the same meaning of cat as a function of repetition of the stimulus word. Subjects report that they characteristically embellish or otherwise change their concept of the stimulus word as a function of repetition. This is understandable as the task is a relatively elementary one, and in a very real sense boring and trivial. Thus, AEPs may be contaminated by the dynamic nature of the semantic considerations of the stimulus word. Hopefully, experimental designs can be developed to at least partially overcome the confounding effects of stimulus repetition.

The third difficulty in neurolinguistic experiments lies in the utilization of concrete words with a referent in the environment. These are words such as 'cat'. Such words possess a built-in bias in that not all individual's conception of 'cat' refers to the same physical stimulus. It is of course possible to limit the considerations on the part of the subject by constructing a model 'cat' and asking that they consider only this definition of the stimulus word. While this may help to reduce the variability somewhat, it does not deal with the central problem, that of the capacity of the subject to form a mental or 'visual' image of the stimulus word itself. Although such studies have not yet been done, it is not inconceivable that the brain response to words that can be easily imaged is different from those words that are not readily amenable to visual imagery (i.e. words denoting the concepts of 'fear', 'freedom', and 'love'). The use of the latter words, however, points up another problem in neurolinguistic research, that of the surplus meaning that each individual brings into the experimental situation regarding the words that have been chosen for study. An alternative strategy is to create words in the laboratory using the familiar nonsense syllables employed in memory studies, defining them as some construct. This, however, is not necessarily a solution, as the whole process is only put through another stage of definition.

There are thus several sources of variability in dealing with neurolinguistic experiments. One source of variability is that inherent in the nature of the brain response to repeated stimulation, even assuming no change in the subjects' cognition of the stimulus. In addition, variability exists between subjects in terms of their experience with particular stimulus words. Furthermore, there may well be variability associated with the anatomical idiosyncracies of individual human brains. It is by no means certain that the neuronal mechanisms for analyzing the word 'cat' are the same from one individual to the next.

The future of neurolinguistic research. While it may seem futile to speculate on the direction that an emerging field may take, there nevertheless seem to be several areas currently being explored in attempts to understand brain correlates of language behavior. One can foresee the continued application of electrophysiological techniques in an effort to map out the language portions of the brain by surface recording of brain potentials in man. It seems relatively unlikely that intrusive electrical measurement of brain processes in man will ever be employed using current technology; however, it is possible that non-intrusive technology can be developed to allow the recording of depth activity in the brain to linguistic as well as to other kinds of stimulation. Given the current constraints on recording electrical activity of the brain, it is reasonable to assume that efforts will continue directed toward an understanding of the brain mechanisms underlying the psychological process of meaning as it is applied in a linguistic sense. It is also reasonable to assume that given suitable progress in the understanding of brain correlates of meaning in linguistic experiments that these techniques and approaches can be applied to other kinds of cognitive performances on the part of human subjects. In addition to the semantic correlates of linguistic behavior as recorded electrophysiologically, it is expected that syntactic linguistic behavior will be investigated as well. It is an open and intriguing question as to the nature of brain responses to certain linguistic stimuli and syntactic relations. While work has not yet been reported relating to syntactic neurolinguistic behavior, it is expected that the future will include such investigations.

Dr. TIMOTHY J. TEYLER, Neurobiology Program, Northeastern Ohio Universities College of Medicine, *Rootstown, OH 44272* (USA). Tel. (216) 678 3004.
Prof. RICHARD F. THOMPSON, Department of Psychobiology, University of California, *Irvine, CA 92664* (USA). Tel. (714) 833 5964.

Language and Hemispheric Specialization in Man: Cerebral ERPs.
Prog. clin. Neurophysiol., vol. 3, Ed. J. E. DESMEDT, pp. 28–35 (Karger, Basel 1977)

A Reconsideration of the Cortical and Subcortical Mechanisms Involved in Speech and Aphasia

HANS H. KORNHUBER

Department of Neurology, University of Ulm, Ulm

Speech has an element of thought and an element of verbal communication which requires the production of voluntary movements. Both the classical and contemporary theories of the cerebral processes of speech have in common that they consider the motor programming underlying speaking to be a cortical process; they disregard subcortical mechanisms involved in the production of voluntary movement. Usually, it is assumed that the flow of information in speech runs from the ear to the temporal lobe, from Wernicke's area to the frontal lobe, and from Broca's area to the motor cortex. According to this view, the cerebellum and the basal ganglia are not supposed to be necessary for speaking since the motor cortex is believed to be sufficient for programming the prodiction of voluntary movement.

I believe that such a view is no longer tenable. The elaborately organized motor cortex is a rather late event in the course of phylogenesis. Most animals move very well without a motor cortex, and even in monkeys most movements are still possible after bilateral extirpation of the precentral motor area or after section of the pyramidal tract [cf. for example KUYPERS, 1973].

Extirpation of the vocalization area of the human precentral motor cortex does not result in aphasia as should be expected if the whole flow of information for speech programming would run from Wernicke's to Broca's area and from Broca's area to the motor cortex.

What is most severely disturbed after lesions of the motor cortex is independent finger movement. This corresponds to the fact that, from the sensory point of view, the motor cortex is to be considered as a somatic (tactile and proprioceptive) association area which in rodents mainly subserves lip and tongue movements and which in primates increases greatly due to the need of advanced tactile and proprioceptive regulation of finger movements for climbing. The eye movements are not represented in the precentral motor cortex because somatosensory regulation would be of little use to them. The role of the motor cortex in speaking is certainly more important than in eye movement. The lips and the tongue are well represented in the motor cortex since lip and tongue movements must be precisely regulated by tactile and proprioceptive messages. Lesions of mouth and tongue areas of the motor cortex relating to phonation result, not in aphasia, but in transient dysarthria [PENFIELD AND RASMUSSEN, 1950].

The motor cortex is not the general basis for voluntary movement. Rather it is a specialized tool of tactile and proprioceptive adjustments for those movements that specifically require this kind of regulation. The general mechanisms of voluntary movement mainly involve subcortical structures such as certain brainstem nuclei, cerebellum, and basal ganglia. Lesions of these structures result in more profound disturbances of voluntary movement than lesions of the motor cortex, as shown by the gaze paralysis after lesions of the pontine reticular formation or the akinesia after basal ganglia lesions [KORNHUBER, 1971].

Technically, a function generator is required to produce a movement, for example, of the electron beam on an oscilloscope screen. For complex movements, one also needs a program generator. Similarly, program and function generators are necessary for the translation of thought into voluntary movements by our brain. The motor cortex is probably not a program or function generator. Without the basal ganglia, cerebellum, and brainstem reticular nuclei, the motor cortex would be ineffective in movement generation [KORNHUBER, 1974a]. As other voluntary movements, speaking requires motor program and function generators.

Why do we have (at least) two cortical speech areas in contrast to one hand area for each hand? To understand this, let us shortly consider the general organization of motor action (to which speaking belongs). Basic functions of actions are to satisfy the needs of the organism and to serve the survival of the species. Therefore, actions must be guided by information from the internal milieu and also from species-related drive information. These are the *strategic aspects* of actions that are represented in the central grey, the hypothalamus and limbic system [see, for example, MACLEAN, 1976] with the cingulate gyrus and frontal lobe cortex as highest levels.

On the other hand, actions must fit into the situation (let us call this the *tactical aspect* of action) and, therefore, they must be regulated by telereceptive messages from the environment that are analyzed in the sensory areas of the occipital and temporal lobe, with the temporal and posterior parietal association areas as highest levels. Strategy and tactics, however, must communicate with each other in order to result in meaningful actions. This is done partly by long association fibers from the temporal and posterior parietal to the frontal lobe (and vice versa), partly by way of thalamic nuclei, for example, from the cingulate gyrus via the pulvinar and lateralis-posterior nuclei to the posterior association cortex. The same communication between strategic and tactical functions is, by the way, also necessary for information reduction between short-term and long-term memory [KORNHUBER, 1973].

Let us now consider the flow of information in the higher processes of speech. The old view of a cortico-cortical arc underlying speech from Wernicke's to Broca's fields is probably less than half of the truth. For one thing, Broca's and Wernicke's fields have direct access to the basal ganglia and (via brainstem) to the cerebellum. Furthermore, the flow of information is certainly not unidirectional from Wernicke's to Broca's area. There must also be influence from Broca's on Wernicke's area since the frontal speech area is juxtaposed and presumably interconnected with frontal areas related to motivational and memory functions (thus, to the strategical aspect of action in my terminology), while Wernicke's area is located near and connected with posterior cortical areas engaged in orientation, perception, and cognition (thus, in the tactical aspect of action).

The organization of language is thus proposed to follow the general scheme of the cooperation of strategic and tactical mechanisms of action. The findings in aphasia are at least compatible with this assumption. Frontal aphasia due to lesions of Broca's area is characterized by lack of fluency, disturbance of syntactic structure, inability to construct whole sentences and agrammatism; this definitely points to a disturbance of the strategy of speech. On the other hand, in Wernickes's aphasia, the incitation to speak is preserved and may even be disinhibited and the formulation of sentences is intact, but lexical formulation is disturbed which results in jargon production.

The anatomical evidence regarding the actual connections of the temporal, posterior parietal, and frontal speech areas of the human brain is rather incomplete. On the other hand, there is little reason to assume that they differ completely from adjacent areas, the connections of which are well investigated in the monkey; and there is some clinical evidence that they indeed agree. According to these data, there are no direct fibers from Wernicke's region to the motor cortex. Broca's and Wernicke's areas are connected by long association fibers in both ways. Broca's area probably projects (indirectly or directly) to the motor cortex. The main pathways, however, from both Wernicke's and Broca's areas to the motor system are direct fibers to the corpus striatum and indirect connections (via brainstem nuclei) to the cerebellum.

Both Broca's and Wernicke's areas have thalamic afferents, Broca's area from the medio-dorsal nucleus (which might transmit information from the motivation system as well as from basal ganglia and cerebellum). Perhaps Broca's area also gets fibers from the ventral anterior nucleus (which has afferents from the pallidum). Wernicke's area receives fibers from the pulvinar-lateralis-posterior complex (which supplies telereceptive and perhaps also motivational information).

Since a distinguished scholar in the field tends to minimize subcortical function in speech [GESCHWIND, 1965], I should like to point out some of the evidence of a thalamic and striatal role in speech. A good review of the data is presented in the first issue [edited by OJEMAN, 1975] of volume 2 of *Brain and Language*. Defects in language and speech have been well documented in cases of thalamic hemorrhage and other vascular thalamic lesions and in stereotactic lesions of the ventro-lateral nucleus as well as the pulvinar.

Disturbances of speech and language occur quite often in the acute stage after left thalamic surgery and are more pronounced after bilateral thalamic (ventro-lateral) or after combined thalamic and pallidal lesions. Ventro-lateral thalamic lesions seem to interfere more often with speech or language than lesions of the pulvinar. Several authors stress that following thalamic lesions there are not only non-aphasic but also higher disorders of speech that are indistinguishable from aphasic disturbances of cortical origin. Disorders of accuracy and completeness of verbal expression, nominal dysphasia and even logorrhoic paraphasia and telegraphic speech have been observed associated with thalamic pathology. However, the most common speech disturbances after left ventro-lateral thalamic lesions are those of fluency and flexibility of verbal formulation and expression, including perseveration. It should be stressed that these effects are not due to involvement of the internal capsula [KRAYENBUEHL et al., 1964] and that they occur in consequence of lesions strictly confined to the ventro-lateral thalamus [SAMRA et al., 1969].

The thalamus itself is, of course, not an origin of speech or language. In my opinion, the general function of the thalamus – which may be deduced from anatomy, namely from the cortical efferents to the thalamus and from clinical phenomena such as thalamic pain – is to regulate the information input to the cerebral cortex [SKINNER and YINGLING, 1977]. What the thalamus may contribute to language or speech comes largely from subcortical structures: from the motivational system and from the subcortical cortical motor system (basal ganglia, cerebellum and some brainstem nuclei).

The cerebellar contribution to speech seems to be of a more simple nature. Cerebellar lesions result in a slow, hesitant or dysarthric speech which agrees with other cerebellar motor symptoms, such as dysmetria, adiadocho-kinesia, and lack of coordination of fast movements. The cerebellum seems to be a device for the generation and adjustment of fast movements that are too quick to be regulated continuously by outer (e. g. visual) stimuli and that therefore must be preprogrammed and adjusted by motor learning. In addition, the cerebellum seems to contain a mechanism for holding those positions reached by the fast movements and for the temporal coordination of quick successive series of movements [KORNHUBER, 1974a, b].

The contribution of the basal ganglia, however, is more sophisticated. This may be concluded from both stimulation and lesion results. There is a type of basal ganglia speech pathology that so far has not received the attention in speech theory it deserves: the speech tics of striatal origin which can be experimentally reproduced by electrical stimulation in the basal ganglia output at the ventro-lateral (ventro-oral) thalamic nucleus. The lit-terature on striatal tics has been summarized [WEINGARTEN, 1968]. Among other tic actions, such as compulsory dancing steps, vocal utterances have often been observed after an encephalitis epidemic, especially compulsive palilalia and coprolalia. The genuine generalized progressive tic disease (Gilles de la Tourette's disease) which is a systematic dysplasia of the striatal cells [BALTHASAR, 1957] is characterized by typical tic movements; in addi-tion, various other extrapyramidal hyperkinesias may occur, such as chorei-form movements. In most cases, the motor and vocal tics are repeated in an individually stereotyped pattern. The vocal expressions, such as offensive oaths and foul expressions, frequently acquire at puberty an obscene, sexual character (coprolalia). In addition, there occurs echolalia, palilalia, and echopraxia in the Gilles de la Tourette's disease. Coprolalia has been reported in cases of chorea [BONNHOEFFER, 1936] which is another striatal disease. Similar evidence comes from stimulation results. Compulsory speech has been evoked by focal electrical stimulation in the ventro-oral nucleus of the thalamus and in Forel's area, both of which conduct the output information of the basal ganglia [SCHALTENBRAND, 1975].

While the psychic obsessive-compulsive symptoms in cases of tic disease can be relieved by stereotaxic coagulation of the rostral intralaminar and the mediodorsal thalamic nuclei, an additional lesion in the inner part of the ventrolateral nuclei (that transmits impulses from the basal ganglia to the vocalization and mouth region of the motor cortex) is required to relieve the Gilles de la Tourette patients from their compulsory verbal expressions [HASSLER and DIECKMANN, 1973].

Antagonists of the neuroleptic drugs such as apomorphin or *d*-amphetamine cause hyperactive behavior in the rat with locomotion, licking, sniffing, and gnawing which is due to the release of basal ganglia mechanisms. By contrast, lesions of the basal ganglia in animals and man lead to akinesia [cf. DEECKE *et al.,* 1977].

Because of these and other findings, it was originally proposed that the basal ganglia are a motor function generator. Since cerebellar learning is necessary for fast movements only, while conventional feedback is sufficient for slow, stimulus-guided movements, it was tentatively concluded that the cerebellum is responsible for the design and adjustment of fast ballistic movements while the generation of smooth, continuous, non-stimulus-dependent movements of voluntary speed was thought to be a function of the basal ganglia, and therefore the term ramp generator was chosen for their function [KORNHUBER, 1971, 1974a]. The compulsory speech phenomena, however, triggered by dysfunction of stimulation of the basal ganglia or their output fibers point to an even higher function of the basal ganglia, namely that of a motor program generator.

Since except for transient dysartria, no major disturbance of speech results from motor cortical lesions, the major part of speech expression is obviously accomplished by the basal ganglia and (to some extent) by the cerebellum. Although there is some speech dominance of the left basal ganglia in right-handed man, the contralateral basal ganglia may compensate for a left-sided defect, such as shown by the more pronounced defects after bilateral lesions.

Because of his inability to generate motor programs by himself, the patient with a lesion in the basal ganglia depends more on external stimuli and also tends to repeat the still existing programs in an inflexible manner. This hyperdependence on external stimuli may lead in the motor domain to the clinically well-known magnet responses such as the tactile magnet response of the athetotic hand or foot or the tactile magnet response of the head in spastic torticollis or torsion dystonia [ASCHOFF and KORNHUBER, 1975]. Frequent repetitions of those actions which are still possible (palipraxia) is a characteristic phenomenon of basal ganglia disease. In such cases, however, those actions that cannot be voluntarily produced by the patient can often still be carried out in response to external stimuli: the basal ganglia patient is hyperdependent on external stimuli and therefore tends to repeat the examiner's actions, i. e. he shows echopraxia.

In my opinion, perseveration, echolalia, and palilalia, recurring utterances and stereotypic verbal behavior are, however, characteristic of a special type of aphasia which is partly due to basal ganglia lesions: *global aphasia*. This type of aphasia is not just a summation of Wernicke's and Broca's aphasia. Global aphasia differs from Broca's aphasia by better syntactical structure and intonation and by more perseveration and recurring utterances [for a recent description, see STACHOWIAK *et al.*, 1977]. But as far as I can see, basal ganglia involvement has not yet been considered as a causal factor in global aphasia. Global aphasia is usually due to thrombosis of the common stem of the middle cerebral artery, and it is known that low branches of the middle cerebral artery (the lateral striate arteries) supply most of the corpus striatum, part of the pallidum and the entire output of the striopallidum, the ansa lenticularis.

Since animal experiments show some degree of specificity of function within the striate body in agreement with the part of cortex projecting to it, and since the functional deficit tends to be more severe when both the cortical and the striatal substrates of the function are damaged, it should be taken into consideration that the most severe kinds of aphasia may be due not just to combined cortical lesions, but to additional lesions of the basal ganglia.

The similarity of the consequences of basal ganglia lesions with aphasic syndromes of cortical origin, especially with Broca's type of aphasia, should make us more careful in distinguishing between aphasic and non-aphasic disorders of speech. Obviously, the generation of motor programs for language expression is less 'external' for language than has often been assumed.

So much for the left hemisphere and the subcortical mechanism. The investigation of commissurotomized patients [SPERRY, 1974] has shown that, in the right-handed man, the 'minor' hemisphere is capable of simple language comprehension. Some results in our laboratory [HEESCHEN and JURGENS, 1977] indicate that the right hemisphere is normally used in specific speech comprehension situations. The right ear dominance in near-threshold dichotic listening holds true only for comprehension from the syntactical context: if comprehension is based on pragmatic-semantic cues, the right ear dominance which is related to its prevalent left hemisphere connections disappears and it may even be reversed. Obviously, the superior abilities of the right hemisphere with respect to spatial relations and non-verbal ideation are utilized for language in appropriate situations [cf. BOGEN, 1969].

When analyzing the cerebral mechanisms of speech along these more comprehensive views, we end up facing difficult problems which relate to human actions in general: many different parts of our brain contribute to our actions. The leading focus often shifts between all the eight lobes of the forebrain. What is guiding this shift is not a superhomunculus nor is it a single superdrive like Freud's libido. The unprejudiced study of human behavior shows that man, like the higher animals, indeed has many drives. Information processing which is the essence of the mind can go on simultaneously in several parts of our brain. Speech is a complex process which requires simultaneous information processing at several levels, such as pragmatic, syntactic, semantic, and phonetic, some of which is carried out in the so-called 'minor' hemisphere. The unity of a human personality is not due to a small part of the left hemisphere cortex, rather it depends on many fiber connections. The (relative) unity of personality is not a language-dependent quality and it is preserved in totally aphasic patients.

Prof. Hans H. Kornhuber, Department of Neurology, University of Ulm, Steinhövelstrasse 9, D–7900 Ulm (FRG). Tel. (0731) 179 21 00.

Language and Hemispheric Specialization in Man: Cerebral ERPs.
Prog. clin. Neurophysiol., vol. 3, Ed. J. E. Desmedt, pp. 36–47 (Karger, Basel 1977)

Semantic Meaning of Words and Average Evoked Potentials[1]

Robert M. Chapman, Henry R. Bragdon, John A. Chapman
and John W. McCrary

Center for Visual Sciences, University of Rochester, Rochester, N.Y., and
Eye Research Foundation, Bethesda, Md.

Among the many exciting prospects of evoked potential research is the
possibility of studying neural activity related to language processes. The
study of language encompasses a wide range of concepts reflecting, in part,
segmentation of interests of researchers and differential levels of analyses
found useful for different purposes. This provides evoked potential research
with many points of entry, many of which are fraught with problems of
definition, control, and measurement.

The research reported here has taken advantage of the large body of
normative data on the semantic meaning of words along empirically derived,
natural language dimensions. Using words as stimuli, we attempt to show
that there are differences in EP measurements that are systematically related
to differences in semantic meaning predicted by the normative data. Our
research has its roots in Osgood's analysis of semantic meaning [Osgood,
1952; Miron and Osgood, 1966; Osgood, 1971] and Begleiter's provocative
EP work [Begleiter et al., 1967, 1969; Begleiter and Platz, 1969].
The work of Osgood and associates on the psychophysics of semantic
meaning is well defined, objectively measured, and widely tested (remaining
valid for more than 23 different language/culture groups). Based on semantic
differential measures on a large number of scales, their work led to the
finding that a good portion of the total variance in judgments of verbal
meaning can be spanned by three underlying dimensions which have been
called evaluation, potency, and activity (E, P, A). Despite deliberate and

[1] Portions of the data in this chapter were previously reported by Chapman (International Symposium on Cerebral Evoked Potentials in Man. Pre-circulated abstracts, pp. 43–45, Presses Universitaires de Bruxelles, Bruxelles 1974).

independent variation in the sampling of scales, of concepts, and of subjects, three dominant and independent (orthogonal) factors have kept reappearing: an evaluative factor (represented by scales such as good-bad, pleasant-unpleasant, and positive-negative); a potency factor (represented by scales such as strong-weak, heavy-light, and hard-soft), and an activity factor (represented by scales such as fast-slow, active-passive, and excitable-calm). Thus, semantic meaning may be specified by the location of a word on OSGOOD's E, P, and A dimensions. The three numbers specified represent the amounts of E, P, and A components in the meaning of any given word.

OSGOOD [1964] noted that these E, P, and A factors are more reactive and affective than they are sensory or discriminatively cognitive. They are dimensions of connotative rather than denotative meaning. The word *mother* denotes 'female parent', but it generally connotes love, care, tenderness, etc. Each of OSGOOD's three dimensions is bipolar. Words may have either positive or negative values on E, P, and A dimensions with neutral values in the middle. These dimensions are orthogonal so that the three values by which semantic meaning is measured are independent. For example, the words *milk* and *money* have, for Americans, similar semantic meaning values (high positive evaluation, medium positive potency, and neutral activity). These values may be contrasted with those for the word *guilt,* which is similar on P and A dimensions, but is extremely negative on the E dimension.

EPs have been studied in relation to words vs. nonsense syllables [SHELBURNE, 1972, 1973], verbal vs. nonverbal stimuli [BUCHSBAUM and FEDIO, 1970], stop-consonant vs. frequency of phonemes [WOOD et al., 1971], speech stimuli [MCADAM and WHITTAKER, 1971; MORRELL and SALAMY, 1971; MORRELL and HUNTINGTON, 1971], meaningfulness of stimuli [MATSUMIYA et al., 1972], and contextual meaning [BROWN et al., 1973; TEYLER et al., 1973]. Most of these studies have sought hemispheric asymmetries. It is natural that the asymmetry question should receive so much attention since the behavioral evidence after unilateral hemispheric impairment or stimulation (lesion, drug, split-brain, etc.) so strongly implicates the left hemisphere in processing language. Furthermore, a strong bias exists for the concept of spatial localization of function in the nervous system and spatial localization concepts translate rather readily into experimental tests. The logic implicit in much of this work is that finding EP effects to 'linguistic' stimuli localized to the left hemisphere would provide evidence that the EP effects are associated with linguistic processing. This logic fits our general premise that covariation of data from neurophysiological and behavioral domains with appropriate controls is the evidence sought. However, if the covariation were more specific, the evidence would be stronger. A left-right difference cannot be very specifically related to linguistic processes, because it is only a simple dichotomy and one which also pertains to many other variables as well, e.g. pitch discrimination, pattern perception, anatomical asymmetries, hemispheric dominances, tongue movements, etc. More specific than hemispheric asymmetry would be, for example, the covariation found along OSGOOD's dimensions of semantic meaning. Not only does such *intralinguistic variation* provide the opportunity to systematically relate EP effects to a systematic conceptual framework, but also the smaller the variations in stimuli and tasks, the less likely that observed EP effects could be due to confounded variables and the more likely that the EP effects are due to semantic processes.

Experimental Design

A large number of words with scores on three semantic meaning dimensions (E, P, A) are available [OSGOOD; HEISE; personal communs]. From these, we selected words which are relatively 'pure' in the sense that they score high or low on one of the dimensions and are relatively neutral on the other two. Thus, six semantic meaning classes (E+, E−, P+, P−, A+, A−) representing the positive and negative extremes of each of the three dimensions were used. Some sample words from E+ and E− classes are given in figure 1, along with their E, P, and A scale values.

The words belonging to these semantic meaning classes were visually presented and the average EPs for these classes were analyzed. The fundamental idea is that the neural components which are common to the words of a given semantic meaning class will appear in the average response, while those aspects of brain processing which are not common will tend to cancel out. This technique may be viewed as an extension of the principle of averaging from the domain of evoked potentials, where it has been shown to be an extremely powerful tool to the domains of stimuli and meaning. The physical parameters of the stimuli (various letters which comprise the words) will vary, of course, from one word to the next and their effects will tend toward the same average EP for the various groups of words. In short, while the background EEG is being averaged to obtain EPs, the physical characteristics of the words are being averaged to control for their effects and the meanings of the words are being averaged to provide a common core of semantic meaning.

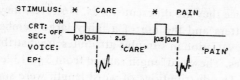

SAMPLE WORDS FROM CLASSES

E+				E−			
WORD	\multicolumn{3}{l}{SCALE VALUES}		WORD	\multicolumn{3}{l}{SCALE VALUES}

WORD	E	P	A	WORD	E	P	A
CARE	1.9	0.6	0	PAIN	-1.8	0.6	-0.1
BATH	1.9	1.0	-0.3	ENVY	-1.3	0.6	-0.1
QUALITY	2.0	1.3	0.6	TROUBLE	-1.2	0.9	0.1
PLEASANT	2.4	0.8	-0.1	CRIMINAL	-1.8	0.1	0.2
FRESH	2.4	1.1	0.3	GUILT	-1.0	1.4	-0.4
BREAD	1.9	0.3	-0.3	ANGER	-1.2	1.0	0.5
PEACE	2.3	1.4	0	DEVIL	-2.0	1.3	-0.4
PERFUME	1.8	0.9	0	FAILURE	-1.5	0.4	-0.3

Fig. 1. Experimental sequence and sample words from evaluative dimension semantic classes. The experimental sequence timing is illustrated for two trials. On each trial, the CRT displays a fixation stimulus (*) and a word stimulus which the subject says aloud. EPs are averaged from the EEG during the 510-msec period following word stimuli belonging to the various semantic classes. Different words are presented on each trial of an experimental run and the sequence of word classes are randomized. Words were assigned to semantic classes on the basis of their scale values on OSGOOD's evaluative (E), potency (P), and activity (A) dimensions which range from + 3.0 to −3.0. The words shown are samples of those placed in E+ and E− classes on the basis of being high or low, respectively, on the evaluative dimension and relatively neutral on the other two semantic dimensions.

20 words from each of the six semantic categories were randomly assigned to a list. Two such lists were constructed with different words, except for the P − category where the same words were used in both lists. The words within each list were given in different random orders from run to run, so that the subjects could not anticipate either a semantic class or a particular word during the experimental runs. Thus, differences in the EPs to these semantic categories can be associated with post-stimulus processing of semantic information, with the comparison of responses to the two lists aiding in the establishment of the reliability and generality of the effects. Because the brain responses to be compared have been derived from semantic categories which are randomly interspersed within each run, it is difficult to attribute any obtained differences to anything other than semantic processing or effects arising from semantic processing.

This conclusion will be reasonable provided that there are no stimulus attributes which are correlated with the semantic categories. The stimuli, of course, are not identical since they are different words. However, they are all composed of the same letters and the use of a fairly large number of words makes it likely that the variation in stimulus attributes are distributed alike to the semantic categories. The word length ranged from 3 to 13 letters with a median of 5.5 letters. The distributions of word length were similar for the semantic classes, with the differences between lists (replication) about the same as the differences between semantic classes.

A computer-controlled CRT display system presents the selected words individually as briefly flashed visual stimuli to the subject seated in a dark, sound-damped chamber. The average word subtended a visual angle of 1.5° with a duration of 17 msec. Each letter was formed by lighting appropriate positions in a 5 by 7 matrix. A fixation target (asterisk) was presented (0.5-sec duration) 1 sec before each word (fig. 1). After each word was flashed, the subject pronounced it aloud (voice in fig. 1) toward the end of the 2.5-sec interval between each word and the fixation stimulus for the next trial. This simple task assured that each stimulus word was perceived. The brain activity following these word stimuli was averaged separately for each of the semantic meaning classes (EP in fig. 1).

The data presented here were recorded from a Grass electrode (silver cup shape) attached by bentonite $CaCl^2$ paste one third of the distance from CZ to PZ (CPZ recorded monopolar to linked ear lobes). The frequency bandpass of the recording system (Grass polygraph, FM tape recorder, operational amplifiers) was 0.1–70 Hz. Beginning with the word stimulus and lasting 510 msec, EPs were averaged by a program using 102 time points (5-msec interval). Each EP was based on 20 different words of the same semantic class. Eye movements were monitored with EOG averages (same gain and frequency response) and EPs from other electrodes were also obtained but not discussed here.

Data from 12 theoretically naive subjects are presented here. The three women and nine men, 18–55 years old, were paid an hourly rate for volunteering. Each subject was given 12 or 18 runs of 120 words (20 words in each of 6 semantic meaning classes) spread over a number of sessions. Half of the runs for each subject were list 1 and the other half were list 2.

EPs to Semantic Word Classes

Our data analysis has concentrated on EPs obtained to words in E+ and E− semantic meaning classes. Average EPs for the first three subjects are shown in figure 2. Examination reveals differences between the responses to E + and E − word classes which are consistent for both lists of words. In order to show the effect on EPs of the E semantic meaning dimension more clearly, the average EP for E − words was subtracted from that for E + words, averaged over three subjects on two word lists, and is shown at the bottom of figure 2.

This average difference based on three subjects was used as a scoring template to measure EPs. A special computer program measured each EP with respect to the scoring template. Using each of the 102 corresponding time points the Pearson product-moment correlation coefficient was com-

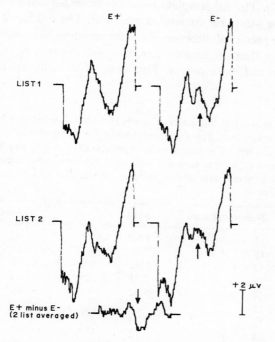

Fig. 2. EPs to words belonging to E + and E− semantic classes (high and low values on the evaluative dimension, respectively, and relatively neutral on the potency and activity dimensions). Each of the two lists contained 20 different words in each semantic class, randomly interspersed. At the bottom is the average difference between the EPs for E+ and E− word classes. This difference vector was used as the E template in measuring evaluative dimension effects in the EPs. EPs are from monopolar EEG recordings between CPZ and linked ear lobes, +up, 510-msec records beginning at time of word presentation. Average EPs for three subjects; n = 360 = 20 words × 6 runs × 3 subjects.

puted. In essence, this was comparable to performing a matched filter. It yielded a single measure for each EP which might then be related to the evaluation dimension of the stimulus words. Using these E template measures, we sought to answer two immediate questions: (1) how sensitive was the procedure in detecting EP differences between E+ and E− classes of words, and (2) what was its generality of applicability (could it be usefully applied to evaluate EPs of other subjects or only those subjects used to develop it)?

The template scoring procedure was applied to EP data of 12 subjects, three involved in developing the template and nine whose EPs had not contributed to the template. The differences between EP measures for E+ and E− word classes were statistically assessed by applying t-tests for each of the 12 subjects (table I). The EP template measures were z-transformed (arc-tanh) before applying t-tests for correlated measures. The t values for all 12 subjects were in the predicted direction, i.e. were positive. The prediction follows from the fact that the scoring template was derived by subtracting E− responses from E+ responses. For many of the subjects the

Table I. The evaluative dimension differences of words (E+ vs.E−) found in brain responses (CPZ) by E template (based on data from three subjects). Number of word stimuli used to obtain each type of averaged evoked response (EP) in each run was 20. Correlation coefficients of scoring template with each EP were z-transformed (arc tanh) before assessing the reliability of the differences between E+ and E− word stimuli by one-tailed t-test

Subject	Number of runs	t-test	Probability
1[1]	12	1.93	<0.05*
2	18	5.25	<0.01**
3	12	1.56	<0.10
4	18	2.05	<0.05*
5[1]	12	3.32	<0.01**
6	18	0.96	<0.20
7	12	0.34	<0.40
8[1]	12	3.12	<0.01**
9	18	3.27	<0.01**
10	18	2.15	<0.05*
11	18	0.35	<0.40
12	18	3.04	<0.01**

Overall: The probability of the number of t-scores meeting the significance criteria by chance is less than 0.001.
[1] Subjects whose data contributed to scoring template.

t-score was sufficiently large to reach statistical significance. In order to combine the findings of the 12 subjects, a method described by SAKODA *et al.* [1954] was used. Including all 12 subjects, the probability of eight cases reaching the 0.05 criterion is much less than 0.001, as is five cases reaching the 0.01 criterion. If the data from the three subjects whose results contributed to the scoring template are removed from the analysis, the probability of five cases out of nine reaching the 0.05 criterion is less than 0.001, as is three cases reaching the 0.01 criterion. Thus, the overall chance probability of the outcomes of all 12 *t*-tests is less than 0.001, supporting the conclusion that differences on the evaluative dimension of visually presented words result in detectable and statistically reliable differences in brain responses. It is to be noted that these template measures were applied to EPs based on relatively small averages (n = 20).

Having found the EP measures to vary in relation to semantic meaning of words, could we compare EP measures and predict backward to what the experimental stimuli were? One way of demonstrating the magnitude of the EP effect is to classify the obtained EPs into E+ or E− classes on the basis of the E template scores. Classifying the larger template measure as resulting from E+ word stimuli and the smaller measure as resulting from E− word

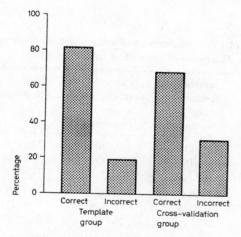

Fig. 3. Percentage of pairs of EPs correctly classified as responses to E+ and E− word classes by the relative magnitudes of correlations with the E template. Each EP was the average response to 20 words. Results for two groups of subjects: template group (three subjects) whose data contributed to the development of the scoring template and cross-validation group (nine subjects). The total number of EP pairs classified was 36 for the template group and 150 for the cross-validation group.

stimuli, we sorted EPs in this fashion. The resulting classification percentages are shown in figure 3. For the three subjects involved in the development of the scoring template, 81% of their EPs (out of 36 pairs) were correctly classified into E+ or E− word stimulus classes on the basis of the relative magnitudes of their correlations with the E template. As expected, a somewhat smaller, but significant, success rate was obtained when sorting the EPs of the nine subjects in the independent cross-validation group. In both groups, the success rates were well beyond what might be expected by randomly classifying the EPs.

The data reported above are for the evaluative dimension of Osgood's semantic analysis. What about the other two semantic dimensions, potency and activity? P and A templates were derived in the same way and the classification results summarized in table II. In general, although the results for the potency and activity dimensions of semantic meaning were not as strong as those for the evaluative dimension, they are encouraging. The effectiveness of the P template in correctly classifying words into P+ or P− classes was sustained in the cross-validation group of subjects. The A template reliably classified EPs into A+ vs A− word classes for the template development group, but was not effective in the cross-validation group.

Table II. Classification of pairs of EPs as responses to + and − words on OSGOOD'S evaluative, potency, and activity dimensions by use of E, P, and A templates

Semantic dimension and EP template		Template development group (3 subjects)	Cross-validation group (9 subjects)	Total (12 subjects)
Evaluative	Correct	29 (81)	103 (69)	132 (71)
	Incorrect	7 (19)	47 (31)	54 (29)
	χ^2	12.2**	20.2*	31.9**
Potency	Correct	23 (64)	106 (71)	129 (69)
	Incorrect	13 (36)	44 (29)	57 (31)
	χ^2	2.2	24.8**	27.1**
Activity	Correct	26 (72)	74 (51)	100 (56)
	Incorrect	10 (28)	70 (49)	80 (44)
	χ^2	6.2*	0.1	2.0

All values of χ^2 corrected for discontinuity. Percentages are given in parentheses. * $p < 0.02$; ** $p < 0.001$.

The relative strengths of the EP effects found for the E, P, and A dimensions of semantic meaning might be expected from OSGOOD's analysis. Evaluation has been found to be the most pervasive aspect of connotative meaning and, statistically, the E dimension accounts for the largest single amount of variation in the connotative values of words, followed by the potency and then the activity dimensions. Furthermore, the semantic distances of our stimulus words between + and − classes from most to least were our E classes (3.2), P classes (2.5), and A classes (1.8). These are semantic distances in terms of mean semantic differential values on the relevant dimensions, full scale being 6.0 (+3 to −3). Our E+ and E− words had mean values of +1.95 and −1.28, respectively, on the E dimension; our P+ and P− words had mean values of +1.91 and −0.61 on the P dimension; and our A+ and A− words had mean values of +0.95 and −0.80 on the A dimension.

Discussion

Evoked potential differences were found between semantic meaning classes using scoring templates based on relatively little data (12 runs on each of three subjects). The EP effects were demonstrated by the success in classifying EPs into high and low word classes according to OSGOOD's evaluative, potency and activity dimensions of semantic meaning. The generality of the particular E and P scoring templates was demonstrated by the classification success achieved in a cross-validation group of nine subjects. The failure to generalize classification success for the activity template beyond the template development group may be related to measured aspects of semantic meaning, as indicated above. Thus, rather detailed and specific covariation between linguistic and EP domains has been found. Differences in EPs were associated with linguistic differences predicted by OSGOOD's quantitative analysis of semantic meaning. This kind of evidence provides strong support to the interpretation that these effects detected in EPs to words are related to linguistic processing.

The use of random presentation order controls for all sources of artifact which might occur prior to the stimulus presentation. This includes eye position, accommodation, pupil diameter, arousal level, background EEG, adaptation level, psychological set, and even whether the subject is performing a specified task or not. Differences between EPs to various semantic classes cannot be attributed to any activity originating before each stimulus word, since the particular semantic category presented on each trial cannot be anticipated due to the random order. Careful examination of the experimental designs in many other EP studies will reveal various pre-stimulus differences which are confounded with the experimental variables of interest. For example, the subject's task may not only be the method of manipulating the linguistic meaning, but also be in force before the stimuli are presented. Another type of pre-stimulus difference in some studies is the prior stimulus which may have different probabilities associated with experimental variables when the research design is attempting to focus on different kinds of discrimination (e.g. stop-consonant vs. frequency discrimination) or context effects (e.g. noun vs. verb meanings). Note that the pre-stimulus differences mentioned here can lurk even in experimental designs which randomize the sequence of the experimental stimuli. For example, noun-verb presentations could be randomized and still have pre-stimulus differences in task or prior stimulus which gives the context. In our experimental design, the differences in EP to semantic meaning classes can be associated with *post-stimulus processing of semantic information.*

Can these EP effects be attributed to attention or arousal? If the differences found between EPs to different semantic categories were due to attention or arousal changes, they would not be generalized reactions, but rather specific ones in the sense that they depend on the various semantic categories and are localized to the time between stimulus presentation and the point where the EP differences occurred. This leads to an important point which has been discussed more fully elsewhere [CHAPMAN, 1973]. Because of the temporal constraint to post-stimulus processes imposed by the randomized experimental design, the acceptance of *any* unidimensional explanation (such general theoretical constructs as alerting reaction, arousal, attention, etc.) permits bracketing the time required for the semantic processes which must occur after the stimulus and before the EP change attributable to the general construct. On the other hand, if we consider the hypothesis that the EP effects are associated with the semantic processes rather than with some general afterproduct, then we can study the processes themselves and their times of occurrence. In the case of this semantic meaning research, it is difficult to stretch any of the unidimensional concepts, like arousal, sufficiently to handle the multidimensional nuances of word meanings found to affect the EPs.

The technique used to measure these EPs was an exploratory one. While based on the use of templates derived from empirical differences, it is not as opportunistic as other procedures that might be applied, e.g. stepwise discriminant analysis. However, it has the virtues of: (1) being relatively direct; (2) being objectively performed by computer, and (3) reducing effects of residual noise in the averages since it involves all the time points. The cross-validation success is a stringent test of the robustness of the semantic effects and their generalizability among subjects. The findings presented here lend encouragement to sharpening the analysis, both in terms of a larger data base and application of a larger range of analytic techniques. The results already strongly support the feasibility of establishing systematic and generalizable relationships between brain and linguistic processes.

Acknowledgments

This investigation was supported in part by NIH Research Grant No. EY 00490 from the National Eye Institute and by NIH Grant No. RR 05698 from the General Research Support Program. Appreciation is expressed to the Neuropsychiatry Division of the Walter Reed Army Institute of Research for their support.

Thanks are due to V. FRATTALI for assistance, to C. E. OSGOOD for helpful suggestions, and to D. R. HEISE for supplying semantic material.

Prof. R. M. CHAPMAN, Center for Visual Sciences, University of Rochester, *Rochester, NY 14627* (USA). Tel. (716) 275 2459.

Language and Hemispheric Specialization in Man: Cerebral ERPs.
Prog. clin. Neurophysiol., vol. 3, Ed. J.E. DESMEDT, pp. 48–59 (Karger, Basel 1977)

Auditory Evoked Potential Asymmetries Related to Word Meaning

RICHARD A. ROEMER and TIMOTHY J. TEYLER

Eastern Pennsylvania Psychiatric Institute, Philadelphia, Pa.,
and Neurobiology Program, Northeastern Ohio Universities College of Medicine,
Rootstown, Ohio

Historically, the evidence pointing to language production and/or integration by Broca's and Wernicke's areas on the dominant hemisphere has come from clinical studies of brain damage [GESCHWIND, 1970; PENFIELD and ROBERTS, 1959]. Electrophysiological analyses of scalp-recorded potentials of these areas in normal humans have emphasized the laterality aspect of evoked potentials (EP) asymmetries to linguistic as opposed to nonlinguistic complex sounds or visual stimuli [BUCHSBAUM and FEDIO, 1970; MCADAM and WHITAKER, 1971; MATSUMIYA et al., 1972; MORRELL and SALAMY, 1971; WOOD et al., 1971; NEVILLE, 1974]. Generally, it has been shown that EPs to linguistic stimuli are of greater magnitude over Broca's and Wernicke's areas of the dominant hemisphere than when recorded concurrently over an analogous area over the non-dominant hemisphere. Recent papers have indicated that, under certain stimulus conditions, differences in EP to the same physical stimulus can be obtained when one of 2 different meanings is attached to the stimulus word, a feature which possibly reflects neurophysiological correlates of 'meaning' [TEYLER et al., 1973; BROWN et al., 1973]. In a previous study [TEYLER et al., 1973], click-EPs to ambiguous linguistic stimuli exhibited hemispheric asymmetries, the dominant hemisphere exhibiting the largest EP. Latencies of the EP also differed, with verb forms of the stimulus word having a shorter onset latency than noun forms. Of considerable interest is the observation that waveform patterns differed between verb and noun forms of the same stimulus word within a given hemisphere. Conversely, the EPs were similar in wave shape over both hemispheres for a given stimulus word-form. In this paper, we report on a more comprehensive analysis of the data collected previously

[TEYLER *et al.*, 1973]. We shall focus on interhemispheric and wave shape differences evoked by the ambiguous linguistic stimuli.

Methods

Subjects were presented with a linguistic stimulus (a word) that could be interpreted one of two ways. For example, the stimulus word 'fire' can be a noun (a bonfire) or a verb (to fire a weapon). In both cases, the physical structure of the stimulus is the same, only the meaning differs. Our approach, then, was to present subjects with several such ambiguous words, and ask them to respond to them in various ways. The word pairs were chosen for their concreteness, ease of imagery and 'distance' between the two meanings to be considered by the subjext [PAIVIO *et al.*, 1968].

Six right-handed coeds were used as subjects. Disc electrodes were placed bilaterally in the region designated C3 and C4 in the 10–20 system. Linked mastoid electrodes were used as the indifferent with an earlobe ground. The EEG was amplified with the band-pass of 0.5–500 Hz and recorded on an FM tape recorder for subsequent analysis. To eliminate artifacts arising from eye movements and eye blinks, surface disc electrodes were placed at the outer margins of each eye and linked together. This signal was amplified, Schmitt-triggered and recorded on the tape as an indication of a source of EEG contamination. The subject sat in a comfortable chair in a sound-attenuating booth and was instructed not to blink or move during a trial. Auditory stimuli were played from a tape and presented bilaterally through head phones. The subject's verbal response, when present, was detected with a microphone and recorded on the tape. The EEG data were averaged in blocks of 10 trials.

The experiment took 2 h and was conducted in 3 phases. In each phase, 3 monosyllabic words were used as stimuli, delivered in random order at intervals of 18–26 sec. Prior to each phase, the subject was instructed as to the meanings of the stimulus words. Stimulus words used in the 3 phases of the experiment were: phase I (1) rock, (2) fix, (3) duck; phase II (1) rock, (2) fly, (3) belt; phase III (1) rock, (2) fire, (3) blow. Note that one world ('rock') appears in all three phases. Phase I consisted of the auditory presentation of the noun or verb form of one of the stimulus pairs, e.g. 'a rock' or 'to rock'. Following a l–to 5–sec pause, the subject heard a click to which she was to respond by saying the stimulus word, e.g. 'rock' while 'thinking of' the meaning that had been given. In other words, in phase I the experimenter presented the subject with a disambiguated stimulus to which she was to respond with the stimulus word. The click also served as a trigger to begin data acquisition.

Phase II consisted of the presentation of a stimulus word, e.g. 'rock' without any modifiers. Thus, the stimulus was ambiguous and could be interpreted as a noun or verb. Following a click, the subject responded by repeating the stimulus word while *thinking of it in one of the two meanings*. After a pause, the subject then verbally reported the meaning she had been 'thinking of' while saying the stimulus word. Thus, in phase II, the subject was allowed to assign a meaning to an ambiguous stimulus.

Phase III was identical to phase II, except that the subject was asked to *not* repeat the stimulus word following the click. Rather, she was only to 'think of' the word upon hearing the click. After a pause, she verbally reported the meaning. Phase III was designed to discern the role of the actual generation of the verbal response in the overall brain response to the situation.

Results

The results to this experiment will be presented both qualitatively and quantitatively. While it would be desirable to discuss only quantitative aspects of the data, current approaches to quantitative analyses of EP fail to account for all facets of the evoked activity. As reported earlier [TEYLER *et al.*, 1973] the ambiguous word paradigm results in hemispheric asymmetries of the major components (N1 to P1) of the EP waveform favoring the dominant hemisphere. Latency determinations indicated no right/left hemisphere differences to the same stimuli, but showed a shorter latency for verb forms as opposed to noun forms of the stimulus word, particularly in the dominant hemisphere. Within-hemisphere waveforms were judged to be different for verb and noun forms of the same stimulus word; whereas between-hemisphere waveforms were similar for a given stimulus word form.

The data reported in the previous study [TEYLER *et al.*, 1973] have been subjected to a further analysis primarily of waveshape characteristics. Figure 1 shows average click-EPs from a subject judged to exhibit the clearest verb/noun differences (fig. 1A) and a subject (fig. 1B) exhibiting the poorest verb/noun differences. Shown are the right and left hemisphere EPs to the verb and noun forms of one word in each phase of the experiment. For each subject, the between-hemisphere EP waveforms are similar. Within-hemisphere differences for verb versus noun are more pronounced for subject A than for subject B. The basic geometry of the waveforms for these 2 subjects consists of an initial negativity, a large amplitude positivity and a subsequent negativity. This triphasic waveform was characteristic of all subjects and all stimuli. Individual differences manifested themselves as perturbations of this basic waveform.

One word, 'rock', was repeated in all 3 phases of the experiment to assess the variability of the evoked waveforms as a function of changing task demands. Figure 2 presents the average right and left hemisphere waveforms to the verb and noun forms of 'rock' across phases for one subject. In contrast to the waveform differences noted in figure 1 to various words in the 3 phases, the data of figure 2 are relatively homogeneous. This result would seem to indicate that the EP waveforms are relatively insensitive to mani-

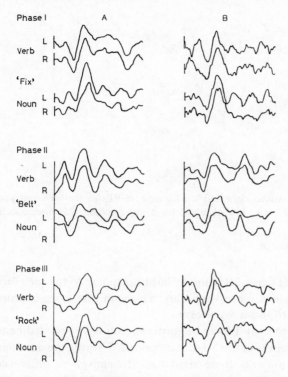

Fig. 1. Average click-EPs from (A) a subject judged to exhibit the clearest verb/noun differences and (B) a subject exhibiting the poorest verb/noun differentiation. Shown are the right and left hemisphere EPs to verb and noun forms of one word in each phase of the experiment. 256-msec epochs.

pulations related to experimenter or subject initiated disambiguating cues or to the subsequent verbal production of the word. These results also indicate that the waveforms remain relatively consistant over the short time span of the experiment.

A characteristic feature of EPs to these linguistic stimuli is that there is great intrasubject variability which effectively limits any attempts at response averaging. Figure 3 graphically depicts this variability for 2 randomly chosen conditions. Figure 3A shows the average right hemisphere responses of 6 subjects to the noun form of 'rock' in phase I. Note that the basic triphasic waveform can be discerned in all the traces, although occurring at different

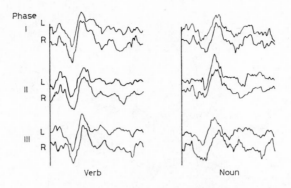

Fig. 2. Average click-evoked right and left hemisphere EPs to the verb and noun forms of the stimulus word 'rock' in each phase for a single subject. 256-msec epochs.

latencies. The tracings of figure 3B from the left hemisphere to the verb form of 'belt' in phase III show more variability which, in one case, obscures even the characteristic triphasic waveform.

Quantitative differences in EPs to linguistic stimuli have revealed hemispheric amplitude differences of the N1-P1 component favoring the dominant hemisphere. Figure 4 presents these right to left hemisphere differences in mean amplitude over all subjects for each word and each phase. The preponderance of difference means favor the dominant hemisphere with the notable exception of the verb form of the third stimulus word in each phase. Two of these words (to 'duck', e.g. avoid being hit; to 'belt', e.g. to hit someone) deal with changing spatial relations of the speaker in relation to objects. It would be appealing to assert that the greater amplitude in the non-dominant hemisphere reflects the processing of this altered spatial relation of the speaker in relation to objects. However, the data cannot support such a statement because other verbs to some degree also imply changing spatial relations (to 'rock' a baby, to 'fly' an airplane) without showing greater non-dominant hemisphere activity. Furthermore, a decidedly non-spatial concept is represented in the non-dominant response to the noun form of 'rock' in phase I.

In an effort to detect commonalities in the EPs across subjects, the data from 6 subjects was averaged for the stimulus word 'rock' which appeared in all phases. Figure 5 (columns 1 and 2) shows these average waveforms for the verb and noun forms of the stimulus word in each phase. As in the

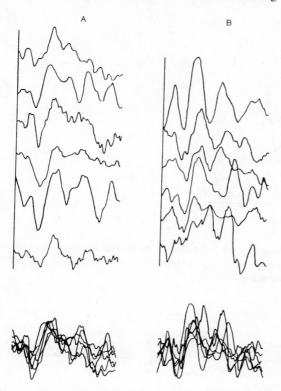

Fig. 3. Variability of the click-evoked waveforms between subjects to the same stimulus. *A* Average right hemisphere EPs of 6 subjects to the noun form of 'rock' in phase I. *B* Average left hemisphere EPs of 6 subjects to the verb form of 'belt' in phase II. Bottom: The same waveforms superimposed. 256-msec epochs.

individual subject records the between-hemisphere waveforms are different for the verb and noun forms of the stimulus word. The pooled waveforms of columns 1 and 2 appear to be relatively similar across phase as they should be since individual records of responses to 'rock' repeated through phases show marked similarities. This apparent visual correspondence was verified by across-phase correlations on the data. Table I shows the result of comparisons made between phase I and II and between phases I and III for hemispheres and word forms. The correlations are high in all cases.

Figure 5, column 3 shows the calculated verb minus noun difference waveforms per hemisphere. Activity above the zero line represents relatively

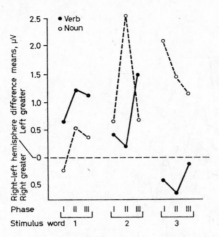

Fig. 4. Mean right/left hemisphere differences in microvolts across subjects for each stimulus word pair in each phase of the experiment. Stimulus word pair number one ('rock') was repeated in each phase.

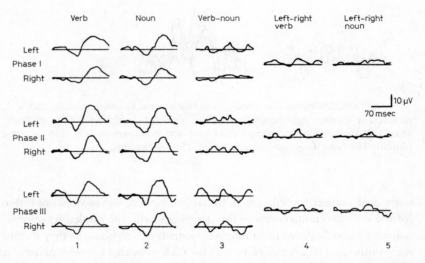

Fig. 5. Average right and left hemisphere click-EPs of 6 subjects to the stimulus word pair 'rock' in each phase of the experiment. Columns 1 and 2: Average waveforms for the verb and noun forms of the stimulus word. Column 3: Verb minus noun difference waveforms (activity above the zero line represents relatively more activity to the verb form). Columns 4 and 5: Right hemisphere minus left hemisphere difference waveforms for verb and noun forms of the stimulus word (activity above the zero line represents relatively more activity in the left hemisphere).

Table I. Across-phase correlations

Phase	Verb		Noun	
	left	right	left	right
I–II	0.90	0.89	0.91	0.92
I–III	0.89	0.86	0.83	0.93

Table II. Verb/noun comparisons by hemisphere

Phase	Left		Right	
	t-score	K-S score	*t*-score	K-S score
I	3.03*	0.1290*	2.99*	0.2234*
II	9.73*	0.1065*	6.12*	0.1047*
III	1.50	0.1063*	0.30	0.1304*

Table III. Left/right hemisphere comparisons by word form

Phase	Verb		Noun	
	t-score	K-S score	*t*-score	K-S score
I	13.30*	0.0778*	15.38*	0.1592*
II	6.94*	0.1465*	5.31*	0.1105*
III	5.79*	0.0794*	1.55	0.1181*

more activity to the verb form, whereas activity below the zero line represents relatively more activity to the noun form of the stimulus word. If the EP waveforms was identical to both verb and noun forms of the stimulus word, the difference waveform would be a straight line (linear function) lying on the zero line (zero mean difference). Waveforms can differ in 3 respects. The difference waveforms can be linear functions with non-zero means, they can be non-linear functions with zero means or they can be non-linear functions with non-zero means. The difference waveforms of column 3 are all non-linear functions with non-zero means.

To test for these linearity and mean differences on the pooled waveforms 2 statistical tests were employed. Linearity tests employed the Kolmogorov-

Smirnov scores of verb/noun difference waveforms by hemisphere. Mean deviations between verb and noun forms were examined with the difference *t*-test. Table II presents the results of these 2 tests for within-hemisphere verb/noun comparisons by phase. The results of these tests indicate that in each comparison there were significant (p<0.05) non-linearities in the verb/ noun difference waveform. Mean differences were also seen (p<0.05) for each comparison except for the activity in phase III. These results suggest that these pooled average waveforms are, in fact, reliably different in phases I and II.

The verb/noun differences between phases I and II can be related to the nature of the task in the 2 phases. In phase II, the subjects were required to decide upon one of 2 possible meanings of the stimulus word. Thus, the larger difference *t*-score may be a task-related effect. Differences in EP magnitude were greater in phase II than in phase III (fig. 4). This result may be due to the elimination of the verbal response in phase III. The effect of not requiring a verbal response implies that motor speech areas would not be utilized and thus motor potentials would not contribute to the evoked activity [GILDEN *et al.*, 1966]. Consequently, the evoked activity in phase III would be reduced relative to phase II. It appears that the mean value of the verb/noun difference scores over phases may be related to motor responses and the other response-related events since the mean does not deviate significantly from zero where there is no motor response requirement (in phase III). It is further possible that the motor related events may tend to mask the effect of meaning since in phase III, where there is no motor response the verb/noun difference waveforms are markedly different from those of phases I and II.

Each of the pooled averages were examined for left/right hemisphere differences in each phase of the experiment. The fourth and fifth columns of figure 5 depict the difference waveforms obtained by subtracting the right hemisphere pooled average evoked activity from that of the left hemisphere for each stimulus word form and phase. In these difference waveforms, above zero activity indicates relatively more activity in the left hemisphere. An inspection of these left/right difference waveforms indicates that there are laterality asymmetries in the pooled averages favoring the dominant hemisphere in each phase of the experiment. Table III presents the results of Kolmogorov-Smirnov and difference *t*-tests on these data. As with the verb/noun differences, there were significant (p<0.05) non-linearities in each difference waveform and significant (p<0.05) mean differences in all comparisons except for the noun form of phase III. Thus, statistical tests indicate

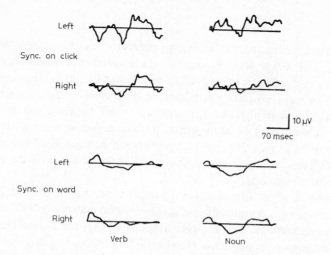

Fig. 6. Average right and left hemisphere EPs to the click and to the linguistic stimulus preceding the click for verb and noun forms of a stimulus word for one subject.

that there are reliable laterality differences in linearity and in mean differences favoring the dominant hemisphere.

The brain activity reported above was all evoked by a click following the auditory presentation of an ambiguous word. Subjects were instructed to think of the priming verbal stimulus upon presentation of the click and, thus, the click can be thought of as a probe stimulus. However, the question remains as to the nature of the EP to the auditory stimulus *per se.* Figure 6 presents the average EPs from one subject to the click and to the auditory stimulus preceding the click. Shown are the average right and left hemisphere activity to the verb and noun forms of the stimulus. The EP obtained to the verbal stimulus follows the characteristics described for the response patterns to click-EPs. That is, within-hemisphere verb/noun differences and between hemisphere similarities for a given word form. It is to be remembered that the verbal stimulus is not an ideal one with which to evoke temporally discrete activity as it has a duration of about 150 msec. Consequently, the waveform does not conform to the basic triphasic nature of the click-evoked activity. Nevertheless, the effect can be observed in verbally evoked brain potentials as has been independently confirmed by BROWN *et al.* [1973] and MARSH and BROWN [this volume].

Discussion

This study has investigated EP waveform characteristics to ambiguous linguistic stimuli. The major findings are that there are marked differences in the waveform of verb and noun forms of an ambiguous linguistic stimulus. These differences in waveform are not related to differential stimulus characteristics as the identical physical stimulus was used to elicit the EPs associated with each stimulus word meaning. Within subject waveforms are variable for a given word meaning, yet group-averaged EPs, while obliterating individual differences, nevertheless, have the same characteristics as observed in individual subjects.

Waveforms for a given stimulus word meaning are highly similar in right and left hemispheres. Amplitude measures of initial components of the evoked waveform do indicate that a larger amplitude response is generally obtained in the dominant hemisphere. However, the observation of similar waveforms in both hemispheres suggests caution in the interpretation of the asymmetric role of the hemispheres in dealing with linguistic information. On the basis of this electrophysiological data, the most parsimonious interpretation is that both hemispheres are involved in some manner in the processing of linguistic information. This is not to say that the processing need be altogether similar in the 2 hemispheres.

In this experiment, all of the stimulus word meanings were specifically chosen to have the attributes of concreteness and ease of imagery. It may be that the electrocortical responses that we are observing are related to the 'mental image' that the stimulus evokes rather than to any other more linguistic attribute of the stimulus. In any event, the EP waveforms appear to reflect the cognitive meaning of the stimulus devoid of its physical attributes.

Several recent reports by others have focused on ERP components of latencies longer than those reported here, and the P300 has been studied in response to one or more words in a short phrase. FRIEDMAN et al. [1975a, and this volume] have observed that each of the words in the phrase elicited a P300, even when the word was predictable and carried no information. The P300 was most reactive to those words that conveyed information regarding the semantic content of the phrase. Thus the cognitive variables normally associated with P300 may be engaged during language behavior. It appears that early and late components may be dissociated, with the early components more sensitive to sensory aspects of linguistic stimuli and late components more responsive to semantic aspects [MEGELA et al., 1977]. Several other reports have dealt with experimental findings [MEGELA and

TEYLER, 1977] and methodological considerations [FRIEDMAN *et al.*, 1975b, and this volume] which are critical of the reports of ERP asymmetries during language behavior.

Acknowledgments

We thank R. F. THOMPSON for facilities and encouragement. Supported in part by Research Scientist Award MH 06650 (R.F. THOMPSON) from the National Institutes of Health; Postdoctoral Fellowship MH 35534 (T.J. TEYLER) from the National Institute of Mental Health; Predoctoral Fellowship MH 48912 (R.A. ROEMER) from the National Institute of Mental Health; and Research Training in Biological Sciences Grant MH 11095 from the National Institute of Mental Health.

Dr. TIMOTHY J. TEYLER, Neurobiology Program, Northeastern Ohio Universities College of Medicine, *Rootstown, OH 44272* (USA). Tel. (216) 678 3004.

Language and Hemispheric Specialization in Man: Cerebral ERPs.
Prog. clin. Neurophysiol., vol. 3, Ed. J.E. DESMEDT, pp. 60–72 (Karger, Basel 1977)

Evoked Potential Correlates of Meaning in the Perception of Language

JAMES T. MARSH and WARREN S. BROWN

Department of Psychiatry and Brain Research Institute, University of California, Los Angeles, Calif.

The investigation of the neural basis of language has until recently been confined primarily to the study of language disorders following lesions of the brain in humans, and many of the neurophysiological methods which can be used with animals are not appropriate for understanding a phenomenon that is uniquely human. This chapter discusses initial experiments in the development of a cerebral evoked potential (EP) method for the study of one linguistic dimension, the contextual meaning of words. Recent studies by TEYLER et al. [1973] and by BROWN et al. [1973] share a common approach to the study of electrophysiological correlates of linguistic meaning in the selection of stimulus words, such as 'fire', 'duck', 'rock', 'fly', etc., which have both noun and verb forms with appreciably different meaning in each form. Thus, differences in cerebral potentials evoked by the 2 forms of such a word can be related to the specific linguistic meaning of the word as perceived by the subject rather than to physical differences in the stimulus. TEYLER et al. [1973] used an expressive paradigm in which the subjects spoke or thought such a word, having in mind a specific meaning. Our studies have utilized a receptive paradigm in which the contextual meaning of the word is provided by a preceding phrase, or instructions to the subject.

Our initial investigation [BROWN et al., 1973] compared potentials evoked by the word 'fire' in the phrases 'ready, aim, fire' and 'sit by the fire'. A second word, 'duck' was used in the phrases 'a flying duck' and 'he'd better duck'. Scalp electrodes were placed over Wernicke's and Broca's areas and over homotopic points on the right hemisphere (linked ears reference). The subject listened through earphones to intermixed repetitions of the prerecorded stimulus phrases. The tapes were prerecorded from tape loops into which the exact same stimulus word 'fire' had been spliced. Thus, the physical parameters of the stimulus word were identical for the two phrases. It was shown that averaged EPs to the same stimulus word differed in waveform as a function of the word meaning. These differences were measured by correlation coefficients computed from the digital values of the average waveform, low correlations indicating dissimilar waveforms. Differences in waveforms were greater (i.e. correlations lower) for left hemisphere recordings than for right hemisphere responses in right-handed subjects. The mean Z-transformed correlation across subjects for the 'fire' experiment was 0.445 for the left anterior locus (Broca's area) and 0.952 for the right anterior recording ($p < 0.05$). Evoked potential waveform differences as a function of contextual meaning were also more different on the left for posterior loci (left mean $Z = 0.561$; right $= 0.836$; $p < 0.10$). Comparing right and left electrode pairs for individual subjects, the left hemisphere correlations were lower than those for the right in 7 of 7 anterior pairs and 8 of 8 posterior pairs. Data for the 'duck' experiment were comparable in direction, but the hemispheric differences in correlations were less marked.

In order to further explore the role of contextual meaning in producing EP waveform differences, a second experiment was performed utilizing the same word 'fire', but this time in the phrases 'fire is hot' and 'fire the gun'. The phrases were presented in two ways to each subject in separate experimental sessions. In one session the phrases were randomly intermixed. In this condition, the meaning of the word 'fire' was ambiguous for the subject during the period of the evoked response, since the subject did not know which phrase would follow until after the EP to 'fire' had occurred. In the other condition, a single phrase was presented 30 times in succession after first informing the subject as to which phrase he would hear. In this condition, the meaning was not ambiguous because there was no uncertainty as to what the following phrase would be. One hundred response to the word fire were averaged for each phrase in each presentation mode (blocked and mixed). As is evident in figure 1, which summarizes the results for 8 subjects, the mean correlations obtained from the blocked presentation of the stimulus are very similar to those obtained in the first experiment, i.e. mean correlations are lower on the left for both anterior (left mean $Z = 0.417$, right $= 0.835$; $p < 0.05$) and posterior leads (left $= 0.621$; right $= 0.811$). In contrast, results from mixed stimulus presentation, where context was ambiguous, showed no differences in mean correlations between the right and left hemisphere (left anterior mean $Z = 0.926$; right anterior $= 0.913$; left posterior $= 0.742$; right posterior $= 0.752$). The increase in correlation over the left hemisphere when context was ambiguous further supports the

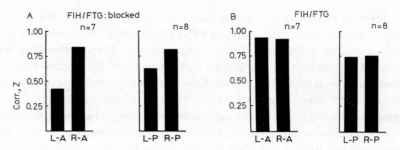

Fig. 1. Mean Z-transformed correlations of average EP waveforms to the word 'fire' in the phrases 'fire is hot' (FIH) and 'fire the gun' (FTG). *A* Correlations are for responses to stimuli presented in blocked form. *B* Correlations are for responses to stimuli presented in randomly intermixed form. L–A=Left anterior electrode placement; R–A=right anterior placement; L–P and R–P=left and right posterior placements.

conclusion that the EP waveform difference was primarily a function of contextual meaning. In addition, the results of this second experiment answered a question raised concerning the 'sit by the fire' 'ready aim fire' experiment, i.e. whether the EP waveform differences to the two contexts could be attributed to differences in the physical properties and temporal relationship of the phrases preceding the stimulus word. Since the phrase followed the stimulus word and stimuli were the same for both blocked and intermixed conditions, waveform differences cannot be attributed to differences in the stimulus phrases. However, these results leave unexplored the somewhat unlikely possibility that in the blocked presentation condition, the physical properties of the preceding stimulus phrase might systematically influence the response to the subsequent word 'fire', while these effects would be averaged out in the intermixed condition. Thus, a different waveform might occur to the word 'fire' when consistently preceded by 'is hot' than when preceded by 'the gun'.

A third experiment was performed which utilized the single ambiguous phrase 'it was led (lead)'. By using a single phrase, all stimulus differences are eliminated. 15 subjects were presented this phrase in four blocks of 60 stimuli. Before each block, the subject was instructed as to which meaning he should attach to the word 'led (lead)', i.e. whether to perceive the word as a verb as in 'the horse was led' or as a noun as in 'the metal was lead'. Results, as seen in figure 2, were similar to those from the first two experiments for the anterior electrode pair (left mean $Z = 0.345$; right = 0.647;

p <0.05). There was little hemispheric difference in mean correlations for the posterior pair (left = 0.659; right = 0.670). Thus, in a situation in which the stimuli were identical and context was provided only by the subject's set to perceive 'led (lead)' as a noun or verb, the waveforms of averaged responses to the two meanings of the stimulus word were markedly different in recordings from the left anterior placement.

Can the lateralized waveform differences obtained in these experiments be attributed entirely to neural events related to the differential processing of meaning or are other factors involved? One possibility is that the low correlations, i.e. dissimilar EP waveforms, obtained in these experiments were produced by trial-to-trial variability either in the response itself or in the background EEG. Thus, differences in correlations would be due to differences in variability between homotopic areas of the two hemispheres. In these experiments, a variance measure was obtained for each average evoked potential. This measure is the sum of the variances calculated from each data point of the average response. Because of the very small signal-to-noise ratio of scalp-recorded auditory EPs to words (e.g. EP amplitudes of 2–4 μV), it seems reasonable to assume that most of the variance measured in this way represents variation in the background EEG in which the EP is embedded. If this form of variance can account for the results obtained, then the greatest variance should be found in EPs from leads which show the lowest correlations. Our data demonstrate the opposite relationship. For each experiment, both the highest correlation between waveforms and the largest variance were obtained from the right posterior lead. Variances in the other 3 recording loci do not differ significantly from each other.

The fact that the experimental results cannot be accounted for by background EEG variability does not necessarily rule out the possibility that the low correlations may be in part a function of trial-to-trial variability of the EP *per se*. This possibility is suggested in our data inasmuch as EPs from the left anterior placement, where the waveform differences are greatest, tend to be smaller in amplitude and have less definitive form than those recorded at other loci (fig. 3A). A method of comparing response variability with the differences in waveform produced by differing experimental conditions has been developed by BUCHSBAUM and FEDIO [1970]. This statistic, called a discrimination index, compares EP correlations from a single locus obtained from two differing experimental conditions with correlations from replications of the same condition. A positive value for the discrimination index indicates that the EPs for the two different conditions were more different than the EPs for the replicates of the same condition. This means

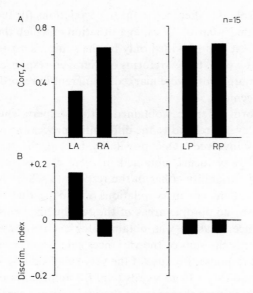

Fig. 2. A Mean Z-transformed correlations for average EP waveforms to the word 'led (lead)' in the phrase 'it was led (lead)'. Bar labels as in figure 1. *B* Mean discrimination index for each of the 4 electrode placements indicated.

that experimental manipulations effect a greater difference in EP waveform than is produced by random response variability.

The discrimination index was computed for the data from the third experiment 'it was led (lead)'. The lower portion of figure 2 shows that the index is positive for the left anterior electrode which shows the lowest mean EP correlation, and slightly negative for the other 3 electrodes. The mean index for the left anterior lead is significantly different from zero ($p < 0.05$). In other words, the EP waveform differences produced by the subject's perception of the word 'led (lead)' as a noun or a verb were significantly greater than the differences in EP waveforms which occur from replication of either the noun or verb conditions.

We have recently submitted the data from the 'it was led (lead)' experiment to two forms of multivariate statistical analysis, Stepwise Discriminant Function Analysis (SWDA) [BROWN *et al.,* 1976] and Principle Component Factor Analysis (PCA) [BROWN and MARSH, in press]. SWDA was performed on the data for 6 individual subjects by dividing the 200 trials into 40 sub-averages of 5 stimuli, 20 for the noun meaning and 20 for the verb meaning. In all 6 subjects, noun and verb EPs from left hemisphere leads could be more accurately discriminated by SWDA than right hemisphere EPs (mean percent correct classification of EPs by SWDA: LA $\overline{X}$ = 79.1%, LP $\overline{X}$ = 76.7%, RA $\overline{X}$ = 71.1%, and RP $\overline{X}$ = 64.3%, 50% being the chance level). In another application of the same procedure, between-subjects SWDA was performed on the final average EPs from all 15 subjects. Here statistically significant discriminant functions could be derived from the EPs from both left hemisphere leads, but not from the right hemisphere data. This latter result is of particular importance in that it indicates a strong degree of intersubject similarity in the EP correlates of contextual meaning. Finally, between-subjects PCA defined six EP components, three of which were shown by an analysis-of-variance of factor scores to be significantly related to the contextual meaning of the stimulus word. Each of these three components appeared to have a unique scalp distribution i.e., there was an early component with a peak at 140 msec which was differentially effected by meaning at both left hemisphere leads, a middle latency component (between 220 and 330 msec) which discriminated between the meanings of the word at the posterior leads on both sides of the head, and a meaning-related late component (385 msec) which was operative only at the left anterior locus. These multivariate analyses are vigorous statistical tests for the reliability of EP differences and thus give strong support for the existence of a reliable contextual meaning-related effect in these data.

We have concluded from the results of the first 3 experiments that the average EP differences observed are related to the neural processing of some aspect of linguistic meaning. If, in fact, the differences do reflect language processing, this relationship should be independent of the modality through which the language stimuli are perceived. Therefore, it should be possible to demonstrate lateralized EP differences to contextual meaning when the same stimuli are presented visually. To test this question, an experiment was performed using the same paradigm as experiment 2, i.e. presenting 'fire is hot' and 'fire the gun' in both blocked and mixed form. 35-mm slides were prepared with the word 'fire' on one slide and either the words 'is hot' or 'the gun' on the next slide. The slides (white letters on black background) were projected through a red filter and the room lights were adjusted to minimize flashed-evoked EP components.

In contrast to the auditory EPs obtained in the previous three experiments, the average EPs from presentation of the slide 'fire' had higher amplitudes, more definite and stable components, and the overall level of correlation between waveforms was higher despite the fact that the visual

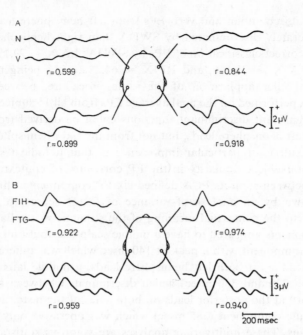

Fig. 3. EPs averaged across all subjects. *A* Across subject average (n=15) of average EPs (n=100) to the word 'led (lead)' in the phrase 'it was led (lead)' presented through earphones. Correlations indicate the degree of waveform similarty of reponses to the noun (N) and verb (V) forms of the stimulus word. *B* Across subject average (n=14) of average EPs (n=50) to the word 'fire' in the phrases 'fire is hot' (FIH) and 'fire the gun' (FTG) presented visually.

EPs represent averages of 50 rather than 100 responses. This is illustrated in figure 3 which compares EPs averaged across subjects to 'it was led (lead)' with responses to 'fire is hot' and 'fire the gun' presented visually in blocked form. (The correlations presented on this figure measure the relationship of the EPs obtained by averaging the data across subjects. They are not the same as the mean Z-transformed correlations in the bar graphs.)

Figure 4 shows that blocked presentation of 'fire is hot' and 'fire the gun' visually produces the predicted differences in mean EP correlations, although the correlation differences between the left and right hemisphere are smaller than those obtained in previous experiments and not statistically significant. However, they are quite consistent as illustrated by the fact that 11 of 14 subjects showed lower correlations on the left for anterior electrode

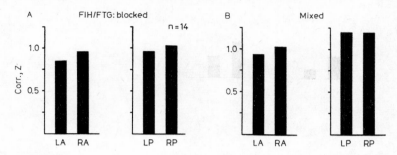

Fig. 4. Mean Z-transformed correlations of average responses to the word 'fire' in the phrases 'fire is hot' (FIH) and 'fire the gun' (FTG) with stimuli presented visually. *A* Correlations of average with stimulus phrases presented in blocked form. *B* Correlations when phrases were randomly intermixed. Bar labels as in figure 1.

pairs ($p < 0.02$) and 10 of 14 for the posterior pairs ($p < 0.06$). Data from the mixed condition, in which the context was ambiguous, showed comparable differences in mean correlation for the anterior pairs, while there was no difference for the posterior pairs. Nine of 14 subjects showed lower correlations on the left for anterior lead pairs, while 6 of 14 showed this difference for posterior pairs.

In contrast to previous experiments, anterior leads showed lower correlations on the left in both the blocked and mixed conditions. In other words, the waveform difference observed at the left frontal lead was greater than that for the right not only when the contextual meaning was clear, but also when it was ambiguous. This suggests a factor operating in both paradigms which is not specific to the differentation of semantic meaning, but is rather a more general correlate of processing language as opposed to the processing of nonverbal information.

Close examination of these data suggest two possible explanations for the differences in results between the auditory and visual experiments. One is that the EP waveform differences related to contextual meaning are specific to the processing of information received by audition. A second alternative is that, with the larger visual EPs, the subtle meaning-related differences in waveform are obscured by large (presumably flash-related) components. This possibility is consistent with the observation that for the visual data EP waveforms are more similar and amplitudes are larger for all leads. Despite the fact that mean correlations showed little contextual meaning effect, at least some trends in the data suggest slight differences between the mixed and

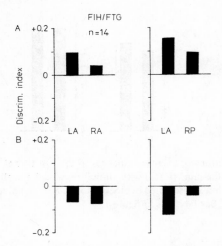

Fig. 5. Mean Discrimination index for the data from 'fire is hot' and 'fire the gun' presented visually. *A* Index for data from blocked presentation of stimuli. *B* Index for data from intermixed condition. Bar labels as in figure1.

blocked conditions. Combining anterior and posterior pairs 21 out of 28 correlations are lower on the left in the blocked condition as opposed to 15 of 28 for the mixed condition. A further indication of possible EP difference present in the blocked but not the mixed condition is found in the discrimination index. Results of this analysis are presented in figure 5. Apparent in this figure is the fact that the index is negative for all recording sites in the mixed condition and positive in the blocked condition. The hemispheric differences in the blocked condition are apparent in both anterior and posterior leads. However, for the anterior pair the mean discrimination index is significantly different from zero only for the left hemisphere (p<0.05), while both posterior leads are significantly different from zero. Despite the fact that EP waveform differences appear to be less well lateralized in this experiment than in the auditory experiments, the discrimination index indicates that the waveform differences are a function of the difference in the conditions, i.e. ambiguity or clarity of meaning, rather than random variability of the EP waveform.

In summary, the data from at least the first 3 experiments using auditory stimulation are quite consistent in demonstrating EP waveform differences which can be interpreted as reflecting differences in the contextual meaning of words. The meaning-related EP waveform differences appear to be most marked in recordings taken from the left hemisphere particularly for the anterior lead located over Broca's area. Summing across experiments for 98 electrode pairs in 50 right-handed subjects, 74% of the pairs showed lower correlations over the left hemisphere for experimental runs. For the control experiments in which meaning was ambiguous, 53% of 60 correlation pairs were lower on the left.

These results suggest that the phenomenon is surprisingly robust in view of all the factors which would seem to mitigate against it. These would include fluctuations of attention over the course of a run, variations in associations and imagery to the same stimulus, the probability that events related to the processing of meaning may not be very precisely time-locked to the stimulus word and, finally, the very low signal-to-noise ratio requiring extensive signal averaging. In light of these factors, it is somewhat surprising that any contextual meaning effect on EP waveform is detectable.

This is not to say that this is not a subtle phenomenon or that there are not problems in eliciting it. For example, the determination of optional sample size for the EP average represents a dilemma. On the one hand, if the sample size is too small, average EPs are not reliable and contextual meaning produced waveform differences do not appear. On the other hand, if the sample size is too large, and the same stimulus is repeated too frequently, semantic satiation occurs and EP waveform differences disappear. We have repeated the same experiment on a single subject on 4 consecutive days. The contextual meaning EP differences resulting from the first day's run were much reduced on the second day, and no differences were observed on the third and fourth days. One possible solution to this dilemma might be the application of such statistical techniques as Weiner filtering of the data which enhances the signal-to-noise ratio allowing for the extraction of an average response from fewer trials.

The fact that a number of linguistic dimensions have been deliberately confounded in the stimulus words selected for these experiments may have a bearing on the degree of robustness of these results. Lexical meaning, part of speech (noun or verb), active or passive nature of the phrase, as well as perhaps word frequency, all contribute to the differences in the contextual meaning of the phrases and, therefore, may all contribute in varying degrees to the EP waveform differences. When the effects of each of these linguistic variables are determined, it may prove that EP differences are too subtle to be detected with respect to any single dimension. On the other hand, the total effect may be a function of just one of these dimensions.

Assuming that the EP differences observed are stable and reliable, what is the source of the waveform differences? In considering this question, we have entertained the following four possibilities:

1) Differences in the contextual meaning of a stimulus word produce differences in the waveform of the potential it evokes as a function of the specific lexical meaning of the word. This would imply differential neural processing for each specific meaning of a particular word which is observable in the EP waveform morphology.

2) Contextual meaning-produced differences in speech EP reflect more general categories of semantic meaning, such as noun or verb, or perhaps categories of word meaning, such as potency, activity and evaluation as measured by the semantic differential scale of OSGOOD et al. [1957].

3) Differences in EPs to different contextual meanings are consequences of evoked response variability from trial to trial. Therefore, instead of evoking specific waveshapes, the processing of contextual meaning increases response variability as compared to situations with ambiguous meaning.

4) Because of low signal-to-noise ratios, increased response variability and consequent EP waveform differences are essentially a function of background EEG. Therefore, during the processing of verbal material, the variability of background EEG must increase selectively over left hemisphere speech areas.

The data that we have collected to date give us some basis for ordering these hypotheses in terms of their probable validity. Thus, hypothesis 4 regarding background EEG would have the least support in our data. As described in an earlier section, we have uniformly found the highest correlations, i.e. most similar waveforms, at loci with the highest variance. In addition, the increases in the level of EP correlations from stimulus conditions in which contextual meaning is evident to those in which it is ambiguous are not accompanied by significant changes in variance. The discrimination index measure computed on the results of the 'it was led (lead)' and visual experiments bears on hypothesis 3 as well as 4. If, for example, the EP differences in the 'led (lead)' experiment are wholly a function of either background variability or EP variability, these processes should operate equally on waveform correlations between replicates of the same meaning as on correlations between different meanings. The positive discrimination index for the left frontal lead indicates that this is not the case and that contextual meaning produces differences in waveform over speech areas which are greater than differences produced by random variation in EP waveform. The near zero and negative discrimination index for the other 3 leads indicates that these correlations represent what little waveform differences are produced by response variability alone. Taken together, the discrimination index and the direct measurement of EP variance suggest that background variability is not an important source of EP differences and response variability, while it may be a factor, does not have a great enough influence on EP differences to account for our results.

Our data are not as yet adequate to clearly select among the remaining hypotheses. In fact, there is some likelihood that the processes subsumed under these hypotheses are both operative. For example, with reference to hypothesis 1, it is entirely conceivable that there are specific neural events related to the perception of the meaning of individual words, but that these processes are too subtle to be reflected in the gross averaged EPs from scalp leads.

There is, however, a source of evidence which indicates that categories of semantic meaning have a measurable effect on EP morphology (hypothesis 2). CHAPMAN et al. [this volume] reported that differences in EP waveforms are related to differences in the dimensions of the semantic differential scale of OSGOOD et al. [1957]. That is, potentials evoked by words clustering at one pole of the evaluation scale were sufficiently different from the responses to words clustering at the other pole of the same scale, so that a specific template could be constructed based on the difference between the average EP to the 2 conditions. This template could be fitted to the data from additional subjects given the same stimulus conditions, to discriminate the responses to the two conditions. This suggest that the linguistic dimensions of the scale produce characteristic EP waveforms across subjects despite physical and lexical differences in stimulus words.

TEYLER et al. [1973] have interpreted the EP waveform differences that they observed as an electrophysiological correlate of what CHOMSKY [1972] has termed 'deep structures' of semantic meaning. In CHOMSKY's view, 'deep structure... relates directly not to the sound but to the meaning' [CHOMSKY, 1972, p. 16]. Certainly, in the 'it was led (lead)' experiment, the same stimulus sound produced different EP waveforms when perceived to carry a different meaning. Thus, we would agree with TEYLER et al. [1973] that the EP effects observed may be related to CHOMSKY's 'deep structures'. An additional possibility from CHOMSKY's theory is that our results represent not deep structures per se, but the operation of the rules for transformation of linguistic information from 'surface structure' (sound) to its specific 'deep structure' (meaning).

Should subsequent experiments confirm our interpretation that the EP differences are related to the contextual meaning of words, and if it can be demonstrated that the differences are systematically related to some set of basic linguistic variables, a promising beginning will have been made in developing one electrophysiological tool for the study of neurolinguistics.

Dr. J.T. MARSH, Department of Psychiatry, Brain Research Institute, University of California, *Los Angeles, CA 90024* (USA). Tel. (213) 825 5784.

Language and Hemispheric Specialization in Man: Cerebral ERPs.
Prog. clin. Neurophysiol., vol. 3, Ed. J.E. DESMEDT, pp. 73–86 (Karger, Basel 1977)

Average Evoked Potentials and Phonetic Processing in Speech Perception

CHARLES C. WOOD

Division of Neuropsychiatry, Walter Reed Army Institute of Research,
Washington, D.C.

Recent theory and experiments have suggested that speech perception may involve distinct auditory and phonetic levels of processing [STUDDERT-KENNEDY, 1975; PISONI, 1971, 1974; WOOD, 1974, 1975a]. The auditory level is assumed to perform a preliminary analysis of the speech signal into a set of nonlinguistic auditory parameters, while the phonetic level is assumed to perform the processes by which abstract phonetic features are extracted. Evidence supporting the auditory-phonetic distinction has come from a variety of psychophysical experiments, including ear advantages in dichotic listening experiments [STUDDERT-KENNEDY and SHANKWEILER, 1970; DAY and BARTLETT, 1972], the 'phoneme boundary effect' in speech discrimination [MATTINGLY et al., 1971; PISONI, 1971, 1974], patterns of interaction between auditory and phonetic dimensions in speeded classification experiments [WOOD, 1974, 1975a], and patterns of phoneme boundary shifts in selective adaptation experiments [EIMAS and CORBIT, 1973; COOPER, in press]. In 1971, we presented average evoked potential (EP) data which also provide support for the auditory-phonetic distinction [WOOD et al., 1971]. The present paper will: (a) summarize results of our more recent EP experiments which bear directly on the auditory-phonetic distinction; (b) present evidence concerning the relation of these EP data to other well-known EEG phenomena, and (c) briefly outline some conceptual and methodological issues associated with these and related experiments which should be considered in future EP investigations of speech perception.

The basic paradigm used in our experiments is a two-alternative choice reaction time task. Subjects are presented a series of synthetic speech stimuli which vary randomly along a given dimension and are required to identify them as rapidly and accurately as possible by pressing one of two response

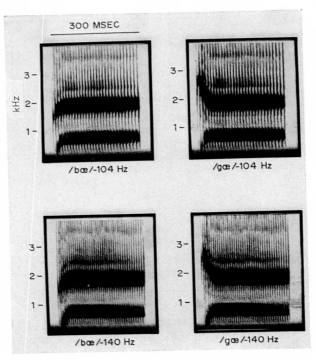

Fig. 1. Spectrograms of the four synthetic consonant-vowel syllables used to compare the processing of auditory and phonetic dimensions. Stimuli which differed along the phonetic dimension (left versus right halves of the figure) differed only in the direction and extent of the second formant transition. Stimuli which differed along the auditory dimension (upper versus lower halves of the figure) differed only in fundamental frequency. All other acoustic parameters of the four stimuli were identical.

buttons. Figure 1 presents spectrograms of the four consonant-vowel syllables used to compare EPs recorded during the processing of a phonetic dimension (place of articulation of the initial consonant) and those recorded during the processing of an auditory dimension (fundamental frequency) of the same stimuli. For convenience, these two dimensions will be referred to as place and pitch, respectively. The stimuli for the place task varied only in the initial consonant (i.e./b/ vs. /g/), with fundamental frequency held constant at one of its two levels (i.e. 104 or 140 Hz). Similarly, the stimuli for the pitch task varied only in fundamental frequency with no variation in the consonant. Following extensive practice in both the place and pitch tasks, each subject received two blocks of 64 trials in each task. The stimulus sequences were constructed so that each of the four possible stimuli shown in figure 1 occurred equally often in both the place and pitch tasks. In this way, EPs during processing of the auditory and phonetic dimensions could be compared without differences between tasks in the acoustic stimuli, their presentation probability, or subjects' motor responses. For a complete description of the procedural details of this and related experiments, see WOOD [1975a].

The EEG was recorded using a Grass Model 7 polygraph with Grass Model P511 preamplifiers. Half amplitude low- and high-frequency settings were 0.1 and 300 Hz, respectively. Recordings were obtained simultaneously from two scalp locations over the left hemisphere (T3 and C3), and two corresponding locations over the right hemisphere (T4 and C4), each referred to a common linked-ear reference electrode. On-line analog-to-digital conversion and signal averaging were accomplished by a LINC computer. All channels of EEG data together with stimulus and response pulses were also recorded on analog magnetic tape for subsequent off-line analyses.

Figure 2 compares EPs at left and right hemisphere locations recorded during processing of the place and pitch dimensions. Each trace in the figure is an average across 12 right-handed subjects, with each subject's average consisting of 128 trials for each dimension. An estimate of the statistical reliability of obtained EP differences was made using Wilcoxon matched-pairs signed-ranks tests at each of the 256 sequential time sample points in each pair of EPs. The results of these analyses are presented in histogram form below each pair of EPs. In addition to providing an approximate estimate of the statistical reliability of the EP differences, this procedure also provides a relatively precise estimate of the distribution of such differences in time relative to stimulus onset and subjects' reaction time responses. However, since the tests at sequential time points are not statistically independent, each pair of EPs was also compared using Wilcoxon tests on conventionally measured base-to-peak amplitudes of the prominent negative component at approximately 120 msec. All significant differences in the point-by-point analyses described below were also significant according to the base-to-peak analyses.

The results shown in figure 2 confirm those originally obtained by WOOD et al. [1971]. At the two locations over the right hemisphere, there were no more significant differences between EPs for the place and pitch tasks than would be expected by chance. In contrast, EPs for the place and

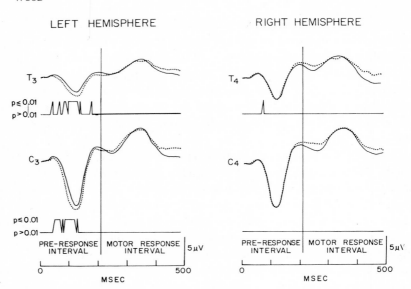

Fig. 2. Average EPs recorded during processing of the place (dotted lines) and pitch (solid lines) dimensions at corresponding scalp locations over the left (T3, C3) and right hemispheres (T4, C4). Positivity at the scalp electrode relative to the linked-ear reference is upward. Upward deflections in the statistical histograms at each location indicate that the difference between EPs at that individual time point was significant at $p<0.01$. From WOOD [1975a].

pitch tasks differed significantly at both left hemisphere locations, beginning between 60 and 80 msec and continuing in time until approximately 120–140 msec. The fact that such differences occurred only at left-hemisphere locations provides additional support for concept of hemispheric specialization of speech perception, derived from analysis of language disorders following brain damage [GESCHWIND, 1970] and the results of dichotic listening experiments [STUDDERT-KENNEDY and SHANKWEILER, 1970].

The relation between the EP results shown in figure 2 and subjects' reaction time performance deserves further consideration. First, it should be noted that the reaction time distributions for the place and pitch tasks were statistically indistinguishable, with means of 416 and 411 msec, respectively. Therefore, it is unlikely that the EP differences were an indirect effect of differences in reaction time between tasks. Also related to this question is the time course of the EP differences relative to subjects' reaction time distributions. The vertical lines at 208 msec in figure 2 represent the time point after which 99% of subjects' reaction times occurred. Thus, the systematic EP differences between tasks occurred in time before subjects began to make their responses. While apparently not a factor in this experiment, the possibility of interactions between movement-related activity and EPs in reaction time tasks is an important interpretative issue and is considered in greater detail below.

The time course of the EP differences between the place and pitch tasks may also have implications for their relation to conventionally defined EP components. As noted above, the statistically significant EP differences began at approximately 60–80 msec and continued to approximately 120–140 msec. This range contrasts with the corresponding significant differences obtained in the Wood et al. [1971] experiment which extended from approximately 100 to 200 msec. Since the peak latencies of the prominent negative component were very similar in the two experiments, the different time course implies that the EP differences may not be directly linked to a particular component as conventionally defined.

In order to investigate which specific aspects of phonetic processing are reflected in the EP differences between auditory and phonetic tasks, another experiment was performed which was identical in all respects to the first except that the synthetic speech stimuli were altered in the following way. Inspection of the stimuli in figure 1 reveals that the syllables /bæ/ and /gæ/ differed only in the direction and extent of the rapidly changing initial portion of the second formant. Formants refer to frequency bands of relatively high acoustic energy and are indicated in figure 1 as dark horizontal bands. The stimuli shown in the figure were prepared with two formants, conventionally labelled F1 and F2 from low to high frequency. Because the syllables /bæ/ and /gæ/ differed only in the initial rapidly changing portion of the second formant (termed the F2 transition), all other parts of the stimuli were identical in each pair of stimuli and therefore provided no acoustic information which could be used to discriminate between /bæ/ and /gæ/. This property of the stimuli was exploited in the design of the second experiment as shown in figure 3.

The left stimulus in figure 3 is the syllable /bæ/ with a fundamental frequency of 104 Hz, identical to that used in the first experiment and shown in the upper left quadrant of figure 1. The right stimulus in figure 3 illustrates how the stimuli for the first experiment were altered. Specifically, all acoustic information which was identical for the syllables /bæ/ and /gæ/ (i.e. which provided no discriminative information between syllables in the first experiment) was eliminated, leaving only the F2 transition. All four stimuli in the

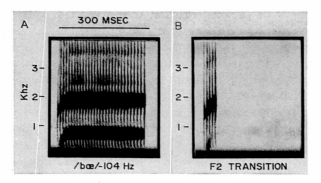

Fig. 3. Spectrograms illustrating the relationship between the full syllable stimuli (A) and the isolated second formant transistions (B). See text for details.

first experiment were systematically altered in precisely the same manner as the syllable shown in figure 3. Phenomenologically, such isolated F2 transitions are typically not perceived as speech, but are heard as nonspeech pitch-glides or 'chirps' [MATTINGLY *et al.,* 1971]. Thus, if the EP differences between the place and pitch tasks in the first experiment were related to phonetic processing *per se,* then no such differences would be expected with the isolated F2 transitions. However, if the EP differences were related to the detection of particular acoustic features in speech, such as the F2 transitions, then results similar to those of the first experiment should be obtained.

EP data recorded during processing of the F2 transition stimuli are shown in figure 4, the format of which is identical to that of figure 2. As can be clearly seen, EP for the pitch and F2 transition tasks differed no more than would be expected by chance at any of the four electrode locations. Base-to-peak comparisons of the prominent negative component were also not significant. These results suggest that the EP differences between place and pitch dimensions were in some way related to the process of phonetic categorization, not to the detection of specific acoustic cues such as the F2 transitions. For a more thorough discussion of the implications of these results for models of phonetic processing, see WOOD [1975a].

More recently, the importance of phonetic categorization in the perceptual differences between syllable stimuli and their isolated acoustic segments has been questioned. For example, WOOD [1976a, b] showed that the 'phoneme boundary effect' occurs for isolated acoustic segments of syllables even though the isolated segments are not categorized by the

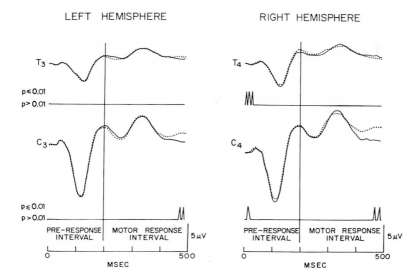

Fig. 4. Average EPs recorded during the processing of pitch (solid lines) and the isolated second formant transitions (dotted lines). Details as in figure 2. From WOOD [1975a].

listeners as speech. These and related results suggest that differences in the perception of a given acoustic segment in syllable context and in isolation [e.g., MATTINGLY *et al.,* 1971; WOOD, 1975a] may be due to differences in acoustic properties of the stimuli, totally apart from the presence or absence of phonetic categorization. Conclusions that perceptual differences between acoustic segments presented in and out of phonetic context directly reflect phonetic categorization may be premature unless discriminability, masking, and other interactive effects can be ruled out.

Although both the WOOD *et al.* [1971] experiment and the experiments just described focused exclusively on average EP analyses of the electrophysiological data, it is also possible that differences between auditory and phonetic processing might be reflected in other measures of electrophysiological activity. This possibility may be related to any or all of the following characteristics of the signal-averaging process as used in our experiments: (a) it does not directly consider the ongoing EEG as a possible measure of differences in activity between auditory and phonetic tasks; (b) it precludes analysis of time intervals other than the specific EP-averaging epoch

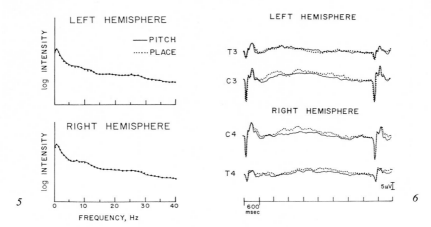

Fig. 5. Mean EEG frequency spectra for left hemisphere (C3) and right hemisphere (C4) locations during the auditory and phonetic tasks.

Fig. 6. Averaged activity at left (T3, C3) and right (T4, C4) hemisphere locations during the inter-stimulus interval in the place (dotted lines) and pitch (solid lines) tasks. Positivity at the scalp electrode is upward.

employed, and (c) by synchronizing on the onset of the acoustic stimuli, this procedure implicitly assumes that the neural activity of interest is also synchronized to stimulus onset. To investigate the degree to which auditory and phonetic processing might be reflected in measures of neural activity other than the average EP, the data recorded in the place-pitch experiment described above were analyzed for differences in background EEG, pre-stimulus baseline shifts related to the CNV, and activity synchronized to subjects' reaction time responses.

In order to determine whether the obtained EP differences were accompanied by differences in background EEG, frequency spectra and amplitude distributions were computed for the EEG data in the place and pitch tasks. For each subject, an average spectrum was computed at one electrode location over each hemisphere (C3 and C4, respectively), by a PDP-12 computer using a Fast Fourier Transform algorithm. The cross-subject averages were then analyzed in exactly the same way as the EP data presented above. Amplitude distributions were computed and analyzed in a similar fashion. Figure 5 presents across-subject averages of the EEG frequency spectra for the left and right hemisphere locations. These data directly correspond to the average EPs at C3 and C4 in figure 2. In contrast to the EP differences which occurred during these same blocks of trials, there were no significant differences in the EEG spectra at either location. A similar pattern of results was obtained in anylyses of EEG amplitude distributions; there were no significant differences in either the mean, variance, skewness or kurtosis of distributions for the place and pitch tasks. Taken together, these results suggest that the EP differences between auditory and phonetic processing were not associated with generalized differences in background EEG.

Since the EP averaging was limited to a 490-msec epoch following the onset of the syllable stimuli, it is possible that the place and pitch tasks also differed in pre-stimulus baseline shifts related to the CNV. This possibility is particularly important in light of the suggestion discussed above that the EP differences may not be directly linked to a particular component as traditionally defined. In order to assess possible differences in pre-stimulus baseline activity, the data from the place-pitch experiment were averaged using a sampling epoch of 6,144 msec, rather than the 490-msec epoch used previously. Since the inter-trial interval between stimuli was 5 sec, this epoch included two successive stimulus presentations plus the inter-trial interval between them. The resulting averages were analyzed statistically in a manner identical to the 490-msec EP data presented in figure 2. The results of this procedure are shown in figure 6, which presents the 6-sec sampling epochs for the place and pitch tasks at each of the four electrode locations. These results provide no clear evidence of differential baseline shifts between the place and pitch tasks. While there appeared to be small positive-negative baseline shifts during the interstimulus intervals, these did not differ significantly between the place and pitch tasks. It should also be noted parenthetically that the decreased sampling rate required for this analysis substantially decreased the temporal resolution of activity following stimulus onset, therefore precluding reliable detection of the EP differences shown in figure 2. Although a tendency toward similar differences at left hemisphere locations can be seen in figure 6, these differences were not statistically significant in the 6,144 msec analyses.

The possibility that the EP differences between auditory and phonetic tasks might be related to differences in subjects' motor response or reaction time was mentioned above, and evidence was presented which argues against such an interpretation. In order to investigate this possibility further, the data were averaged with synchronization on subjects' reaction time response rather than on stimulus onset as in the previous analyses. A

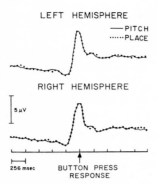

Fig. 7. Averaged activity at left (C3) and right (C4) hemisphere locations synchronized to subjects' reaction time responses. The point of synchronization is indicated by the
arrow. Positivity at the scalp electrode is upward.

LINC computer sampled and averaged data 1,280 msec before and after each
reaction time response, resulting in the data shown in figure 7. This figure
presents across-subject averages synchronized to subjects' reaction time
responses at one location over the left hemisphere (C3) and one location over
the right hemisphere (C4) in both the place and pitch tasks. The statistical
analyses indicated that potentials for the place and pitch tasks were indistinguishable. Thus, these results support the suggestion that the obtained EP
differences between auditory and phonetic tasks were not associated with
differences in subjects' motor response. In addition, these data imply that
the differences between tasks are better synchronized in time to the onset
of the acoustic stimuli than to subjects' reaction time responses.

Taken together, the data presented above support the interpretation
that different EP activity occurred during the processing of auditory and
phonetic dimensions of the same speech stimuli. Such differences were
localized in time to the interval between stimulus onset and subjects'
reaction time responses, and did not occur during the processing of isolated
F2 transitions. The following paragraphs outline some conceptual and
methodological issues associated with these and related experiments which
should be considered in future EP investigations of speech perception.

The experimental design used to obtain the data presented above is
somewhat unusual in that EPs were recorded under conditions designed to
differ only in the information-processing tasks subjects were required to

perform. Thus, these experiments attempted to control as many other variables as possible and isolate differences in processing as the primary independent variable between tasks. An important advantage of this design is that identical stimuli were used in the two tasks, therefore precluding possible EP differences related to stimulus variables, such as spectral structure, amplitude, amplitude contour, rise-time, and duration. Such stimulus variables are confounded with processing variables in experimental designs which compare EPs to speech and nonspeech stimuli.

A second advantage of the 'identical-stimuli-variable-task' design is that EPs for the two tasks can be compared from precisely the same electrode locations. In contrast to EP comparisons between different electrode locations, within-location comparisons eliminate the possibility that EP differences could result from variations in electrode placement, differences in amplification characteristics of the recording apparatus, or anatomical differences associated with different electrode sites.

Comparing EPs for different tasks at the same electrode location provides a more subtle advantage in that it allows a very precise null hypothesis to be stated *a priori*. Such logical precision is impossible using comparisons of EPs derived from different electrode locations. Under conditions in which the null hypothesis is true (i.e. no true difference in EPs for different processing tasks), the formal expected value of the difference between EPs at a single electrode location is zero. In other words, when the null hypothesis is true, EPs derived from a single electrode location are random samples from the same population and should differ only to the extent expected by chance variability. This is exactly the pattern of results described above for the full syllable stimuli at right hemisphere locations and for the F2 transition stimuli at locations over both hemispheres. In contrast, a precise expected value cannot be stated *a priori* for differences between EPs derived from different electrode locations, since slight variability in electrode location can produce relatively large differences in EP amplitude and waveform. This problem is particularly important in comparisons of auditory EPs derived from locations over the left and right hemispheres, since such potentials undergo rather sharp amplitude changes in these scalp regions [VAUGHAN and RITTER, 1970; GOFF et al., 1969; PERONNET and MICHEL, 1977].

Both the advantages just discussed and the empirical success of the 'identical-stimuli-variable-task' design in the experiments described above provide strong support for its utility in EP experiments. However, this design involves a number of assumptions which should not be overlooked. Some of such assumptions are associated specifically with this particular design, while others are implicit in virtually all EP investigations of cognitive processes.

The most important assumption made by holding stimulus variables constant and comparing EPs from different processing tasks is that subjects are capable of performing, and in fact actually perform, the specific information-processing functions intended by the experimenter. Although this assumption appears to be trivial at first glance, recent psychophysical evidence concerning the flexibility of human information-processing capability requires that it be given careful consideration. An example related to the auditory and phonetic tasks described above will serve to illustrate the importance of this assumption.

In order for a comparison of EPs during the processing of auditory and phonetic dimensions to be meaningful, the psychological and physiological processes required to extract auditory and phonetic information from a speech signal must differ in a nontrivial way. Evidence for psychological differences between auditory and phonetic processing is available from a variety of psychophysical experiments, including those cited in the introductory paragraph of this paper. Most directly relevant are the reaction time experiments of DAY and WOOD [1972] and WOOD [1974, 1975a], which suggest that subjects are capable of processing the pitch dimension selectively while ignoring or filtering the place dimension. However, the data of WOOD [1974, 1975b] demonstrate that subjects are also capable of some form of shared, simultaneous or parallel processing of information about the place and pitch dimensions. This capability for optional processing of various forms of stimulus information has been documented in a variety of other experimental contexts [GARNER, 1974; NICKERSON, 1971; NORMAN and BOBROW, 1975; CRAIK and LOCKHART, 1972] and presents a significant interpretative problem for EP experiments [cf. SUTTON, 1969].

In particular, the capability for optional processing requires that EP investigations of cognitive processes either: (a) assume by definition that subjects employ a single processing option; (b) attempt to force or constrain subjects to use a single option, or (c) obtain evidence independent of the EP data of interest concerning exactly what processes are actually employed in a given experimental condition. The first of these alternatives simply ignores the optional processing problem and can therefore lead to serious misinterpretation. The second alternative is currently being employed successfully in selective attention experiments by using stimuli and tasks which exceed subjects' processing capacity and therefore impose a specific processing option [e.g., DEBECKER and DESMEDT, 1970, 1971; HILLYARD et al., 1973].

However, this approach is not widely applicable, since the range of situations in which such clear constraints upon subjects' options can be imposed is relatively limited.

The third approach, obtaining independent evidence of process options actually employed, appears to offer the most general approach to the optional processing problem. This approach is a special case of the need to obtain independent evidence of subjects' performance in EP investigations of cognitive processes. Although the value of simultaneous psychophysical and electrophysiological measures is self-evident, only recently has this approach been systematically exploited in EP experiments. For example, a number of recent experiments have demonstrated the utility of relating detectability and response bias measures derived from signal detection theory to EP recorded simultaneously in the same tasks [HILLYARD et al., 1971; PAUL and SUTTON, 1972; SQUIRES et al., 1973]. Similarly, reaction time measures have become increasingly important tools for investigating a wide variety of human information-processing characteristics [STERNBERG, 1969; PACHELLA, 1974; THIEOS, 1974] and might therefore be useful as concomitant measures of performance in EP experiments. As currently employed, however, reaction time methods present an additional interpretative problem since movement-related activity synchronized to subjects' reaction time responses may influence the EP data. This problem was avoided in the present experiments by matching the discriminabilities of the stimulus dimensions employed so that reaction times (and therefore the temporal distribution of movement-related activity) would be as equal as possible for each task. However, this solution has limited generality since in many experiments the EP comparisons of greatest interest would be between conditions in which reaction times differ significantly. A possible alternative solution to the problem of movement-related activity in reaction time experiments may lie in the use of speed-accuracy trade-off functions [LAPPIN and DISCH, 1972; PACHELLA, 1974; WOOD and JENNINGS, 1976]. This technique would permit reaction time differences to be artificially controlled, without influencing differences in processing between conditions. If successful, the speed-accuracy trade-off approach would make a wide variety of information-processing experiments amenable to EP analyses, experiments which are now potentially confounded by differences in the temporal distribution of movement-related activity.

In summary, the EP data presented here provide additional support for the distinction between auditory and phonetic processes in speech perception previously derived from psychophysical data. The contrasting results of the experiments with full syllable stimuli and isolated F2 transitions indicate that the same acoustic events were perceived differently when they occurred in isolation and when they occurred in phonetic context. The results of the F2 transition experiment suggest further that the EP differences between auditory and phonetic processing are related in some way to the process of phonetic categorization, not to the detection of specific acoustic events such as formant transitions. This interpretation must be tempered by the more recent psychophysical results discussed above [WOOD, 1976a, b] which question the role of phonetic categorization in perceptual differences between acoustic segments presented in syllable context and in isolation. Finally, analyses of background EEG, prestimulus baseline shifts, and movement-related activity suggested that the EP differences between auditory and phonetic processing were localized in time to the interval between stimulus onset and subjects' reaction time responses. A number of conceptual and methodological issues were raised which should be considered in future EP investigations of speech perception.

Acknowledgments

The experiments described in this paper were conducted at the Neuropsychology Laboratory, VA Hospital, West Haven, Conn. They were supported in part by NIMH Grant M-05286 and NSF Grants GB-3919 and GB-5782 to W. R. Goff, and by NICHD Grant HD-01994 to the Haskins Laboratories, New Haven, Conn. The advice and encouragement of W. R. GOFF, R. S. DAY, and W. R. GARNER are gratefully acknowledged.

Dr. C.C. WOOD, Neuropsychology Laboratory, VA Hospital, *West Haven, CT 06516* (USA). Tel. (203) 932 5711.

Cerebral Potentials Associated with Speech Production

Language and Hemispheric Specialization in Man: Cerebral ERPs.
Prog. clin. Neurophysiol., vol. 3, Ed. J. E. DESMEDT, pp. 87–103 (Karger, Basel 1977)

Human Cerebral Potentials Preceding Speech Production, Phonation, and Movements of the Mouth and Tongue, with Reference to Respiratory and Extracerebral Potentials

B. GRÖZINGER, H. H. KORNHUBER and J. KRIEBEL

Department of Neurology, University of Ulm, Ulm

When compared to potentials preceding finger or eye movements, averaged electroencephalographic recordings associated with speech production show greater variability. Some of the potentials which seemed to precede the onset of speech [ERTL and SCHAEFER 1967, 1969; SCHAFER, 1967; MCADAM and WHITAKER, 1971] were later shown to actually follow the start of activity in the speech muscles [GRÖZINGER et al., 1972, 1975]. The determination of the real time of onset of movement is one of the most important and difficult tasks in any investigation of the cerebral correlates of voluntary movement. Another major source of error results from the fact that there is a coordination between the onset of speech and the phase of the respiratory cycle, the reason being that respiration is associated with both cerebral and extra-cerebral potentials which can thus interfere with the speech-related potentials [GRÖZINGER et al., 1973, 1974a, b, 1976]. Averaging is quite efficient to reveal any electrical signals if these are synchronized with the process studied. This type of contamination of the cerebral potentials is difficult to control in the case of muscle activities located in the head itself, as in speech or phonation. The elimination of artefacts due to such sources requires many control procedures.

After a period of enthusiastic over-interpretation of ERPs associated with voluntary movements [KURTZBERG and VAUGHAN, 1973, 1977] there is now overcritical scepticism. At the Marseille and Brussels meetings, papers were presented which deny the cerebral origin of all cranial potentials associated with speech [GRABOW and ELLIOTT, 1973; SZIRTES and VAUGHAN, this volume]. In this paper, we discuss some necessary controls and possible misinterpretations and document several kinds of prespeech potentials of cerebral origin.

Methods

Cranial potentials and EMG preceding speech, articulatory movements and phonation (humming) were investigated in 109 experiments using 32 normal subjects. Handedness was determined by dexterity tests and questionnaires. Bilateral recordings were taken from several scalp electrodes including the vertex, Broca's and Wernicke's regions. For monopolar recordings, linked ear lobes were used as reference; extracephalic locations (sternum and vertebra prominens balanced as to minimize electrocardiogram contamination were used for comparison. Nasal respiration was recorded by means of a thermocouple. For each kind of activity, the earliest onset of movement was determined by means of multiple electromyography recordings (including needle EMG from the diaphragm, fig. 1). Care was taken to trigger the averaging by the real onset of movement. This was for some articulatory movements and speech productions the orbicularis oris electromyogram, for humming or phonation without articulatory movements the larynx activity (phonogram onset) picked up by a throat microphone. For sharp triggering the subjects were required to pronounce the first phoneme of the word as rapidly as possible. Words starting with labials (i.e. Puppe) or dentals (i.e. Lotte) gave sharper triggering than gutturals. For control of facial and tongue movements, electromyograms were obtained from various facial sites. Galvanic skin resistance was measured from the hand. Additional control recordings were taken from the thorax and extremities. Air conditioning was used to reduce galvanic skin potentials. Control experiments were done on abraided skin. Vigilance was checked with electroencephalographic recording and television monitoring of the subject. The experiments were performed under normal room illumination. The eyes were open, fixating on a dot straight ahead. For further details of control procedures, see GRÖZINGER et al. [1975]. While recording on-line, the traces of every trial were controlled by visual observation and marked with positive or negative impulses on the tape after each movement: negative impulses were used if eye blinks or other artefacts occurred. Artefact-free trials were marked with positive pulses which implied acceptance for off-line analysis. The method of analysis was time reversed averaging [KORNHUBER and DEECKE, 1964, 1965] of the movement related potentials using a PDP-12 computer [DEECKE et al., 1969; BECKER et al., 1972; GRÖZINGER et al., 1975]. The sweep duration was either 4 or 8 sec, thus allowing recognition of respiratory potentials. No stimuli were given. The subjects were asked to speak at irregular intervals with intertrial intervals of about 10 sec to avoid overlap of after-effects from preceding trials. The actual intertrial intervals ranged from 6 to 20 sec.

For recording purposes a time constant of 1.2 sec was used for most experiments. The high frequencies were cut off by a filter at 70 Hz. Five control experiments were done with DC recording. In all experiments, DC shifts in the inter-trial interval were eliminated by normalization in the time span between 3.0 and 2.0 sec prior to activity onset. The average brain potentials were calculated from the numbers of trials performed, which varied from 30 to 300 (mean 250). For control of stationarity, the potentials were also analyzed in successive subgroups of 30–50 trials.

Results

Most cerebral potentials were obtained from sentences and words starting with 'L' or 'P' and from utterances of the phonemes 'T', 'B', 'L' and 'A' as pronounced in German. The intersubject variability of the cerebral potentials was higher than the intra-subject variability. Some subjects showed a high degree of reproducibility. In contrast to SCHAEFER [1967] and to SZIRTES and VAUGHAN [this volume] no correlation was found between the structure of particular utterances and the morphology of the pre-speech slow potentials.

1. Importance of the Determination of the Real Onset of Movement
Multichannel recordings of the activity of different muscles show that the use of a microphone as a trigger reveals the activity too late (up to 350 msec after beginning of articulatory movements) for the determination of speech onset (fig. 1). A voice microphone [as used in the studies of ERTL and SCHAEFER, 1967; McADAM and WHITAKER, 1971; MORRELL and HUNTINGTON, 1972; SZIRTES and VAUGHAN, 1973] responds even later, so that cerebral potentials which appear to precede muscle activity by more than 300 msec are in fact potentials evoked by afferent discharge [fig. 2 of GRÖZINGER *et al.*, 1975]. Under our experimental conditions the soft palate muscles were the only ones where movement occasionally preceded that of the orbicularis oris muscle (fig. 1). Myogenic potentials of the soft palate could not account for the potentials recorded at the scalp however since they did not even appear at facial sites. Sensory evoked potentials from the palate should be larger over the postcentral area than over Broca's field.

2. Interhemispheric Differences, the R-Wave, the Bereitschaftspotential (BP), the Galvanic Skin Response and Glossokinetic Potentials
Differences between the potentials recorded over the right and left hemispheres occur earlier preceding language production or phonation of letters than preceding eye or finger movements. Such hemispheric asymmetries are sometimes found as early as 2–3 sec prior to speech onset. We did not directly investigate asymmetries prior to 3 sec because of our normalization procedure between 3 and 2 sec before activity onset (see methods). However, when analyzing the respiratory cerebral potential with an 8 sec sweep time, we have seen marked hemispheric differences as early as 7 sec prior to activity onset (fig. 2).

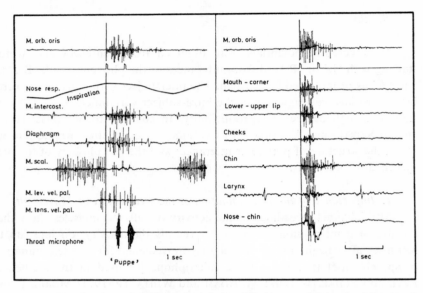

Fig. 1. Coordination of respiration and the different muscles involved in speech activity ('Puppe'=doll). The vertical line indicates onset of orbicularis oris electromyogram. There is latency of 230 msec between the mouth EMG and phonation (throat microphone). The EMG's in the left half of the figure (subject K) were recorded via needle electrodes (except for the orbic. oris muscle); the EMGs in the right half (subject J) were recorded via skin electrodes. Chin and larynx are unipolar leads vs. linked ears, the remaining EMGs are bipolar recordings (VOK 58, K8).

It is impossible to analyze the BP preceding speech or vocalization without critical reference to the respiratory potentials. As has been shown previously [GRÖZINGER *et al.,* 1972, 1974a, b] there are genuine cerebral potentials related to respiration, the so-called R-waves, which often disclose hemispheric asymmetries.

The window in figure 3 indicates that analysis with a short sweep time (as used by most authors in this field) is highly misleading for evaluation of cerebral potentials preceding speech. As revealed by a longer analysis period, the strong positive potential in the last second before onset of mouth electromyogram is, in fact, due to respiration.

The R-wave amplitude is sometimes variable in the same individual and even more so in interindividual comparisons. This might be due to the multiple components of the R-wave (e.g. efferent respiratory activity of the brain, afferent respiratory discharges from mechanoreceptors of the skin,

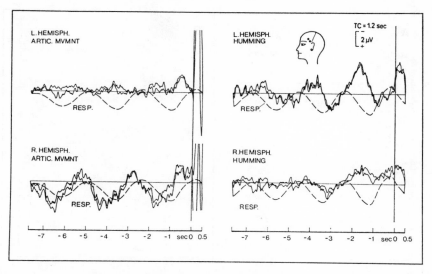

Fig. 2. Averaged brain potentials (solid curves) and respiration pattern (nose thermot couple, broken curves) preceding articulatory movements (left half) and humming (righ-half). Over one hemisphere there is a clear correlation between the average respiratory curve and the wave-like brain potentials (R-waves) whereas the opposite hemisphere almost shows no correlation. In this example, the R-wave appears over the right hemisphere preceding articulatory movements and over the left hemisphere prior to humming. Vertical lines indicate the onset of activity. Sweep time 8 sec, n = 250. Reference linked ears (VOK 56).

muscles, joints and lungs as well as potentials from the lungs themselves) which might have various phase relationships resulting in either summation or substraction, i.e. the R-wave recorded from the scalp might be either enhanced, diminished or even abolished [GRÖZINGER 1974a, b]. The same variability of the R-wave amplitude has also been seen in rats with epicortical recordings [CASPERS, personal commun.]. R-waves have also been recorded from humans at rest without speech activity in our control experiments, using the onset of expiration as a trigger, and from the skull of cats in barbiturate anaesthesia.

In 5 of our 32 subjects there was a polarity reversal of the R-wave between both hemispheres. An example is given in figure 4, showing four different experiments on different days with identical electrode placement. These tracings give some idea of the intrasubject variability of the phenomenon. Preceding finger movement, the BP has an average duration of 0.85 sec

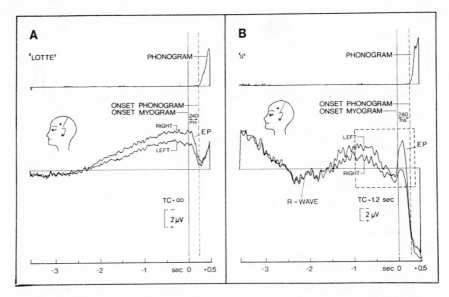

Fig. 3. Showing how misleading a short sweep time (window) would be. The potentials preceding speech onset in B are mainly due to respiration as the long analysis time shows. Short analysis would suggest a positive potential change in the last second before activity onset. By contrast, in A a short analysis time would yield no potential change, while really there is a strong negative potential change starting about 2½ sec prior to EMG activity. Furthermore, the large potential following mouth myogram onset would in both cases be mistaken as a pre-speech potential if phonogram onset had been used for triggering analysis. Above: rectified mouth phonogram. Vertical solid lines indicate onset of orbicularis oris electromyogram. Sweep time 4 sec. Two subjects, n = 230 and 250. Reference linked ears (VOK 40 and 13).

[DEECKE *et al.,* 1969] while preceding speech production and vocalization of single tones, the differences between potentials over both hemispheres certainly show an earlier increase on the average. This is also clear in bipolar recordings. Typical examples of monopolar recorded potentials are shown in figures 5 and 6, and of bipolar recorded potentials in figures 7–9.

Due to the extensive control experiments, the topographic distribution of the cerebral potentials has up to now only been investigated in a relatively small sample.

However, it seems that with single word production, the pre-speech cerebral potentials are more pronounced over Broca's area than Wernicke's area. With production of full sentences, the amplitude difference between

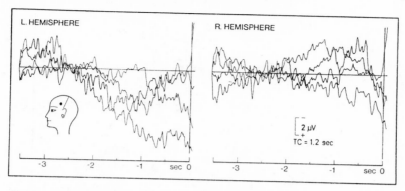

Fig. 4. Averaged cerebral potentials preceding different phonemes and words (in one experiment 'a', in the other the word 'Lotte' and in the third and fourth experiment the word 'Puppe'). Four different experiments with the same person and with the same electrode positions (Broca's field and contralateral homolugue vs. linked ears). In every experiment there were strong R-waves, which are different over both hemispheres. In addition, there is trend towards positivity in the last half second prior to mouth EMG activity (vertical line). n = 180–250 (VOK 22, 30, 57 and 78).

these two is less. An example of a negative potential larger over the left frontal (Broca) area than over the left posterior temporal region is shown in figure 6. In this experiment, the subjects were required to produce about 200 different sentences each starting with the word 'Puppe' (doll). The BP preceding finger movements is negative over the precentral and parietal cortex and usually positive over frontal areas [DEECKE *et al.,* 1973, 1976] whereas, by contrast, the corresponding potential preceding speech and articulatory movements may be either negative or positive at frontal, precentral or temporal sites (fig. 5–7).

To differentiate cerebral from extracerebral potentials, most of our experiments were additionally analyzed in successive subgroups. In most experiments, the shape of the cerebral potentials throughout the experiment remains remarkably constant as shown in figure 10. However, this does not hold true for the R-waves which may change their frequencies or phase relative to the onset of speech. The influence of changes in vigilance as judged by the alpha rhythm and the amplitudes of the average cerebral potentials is small. In contrast, the phase relation of the R-wave to speech onset changes remarkably during the experiments and also the shape and phase of the galvanic skin response, if present.

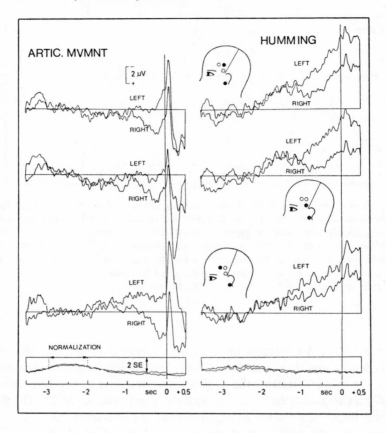

Fig. 5. Two different experiments using the same subject and identical electrode positions on the same morning. Left: Articulatory movements of the mouth (similar to pursing ones lips); right: humming. On the right, there is a clear superimposition of R-waves which are similar over both hemispheres in this case, but less pronounced at the frontal electrode. In addition, there is a marked interhemispheric asymmetry in the last 1½ sec prior to activity onset. In the case of articulatory movement without phonation (left), there is no coordination with the phase of respiration and, therefore, less R-wave apparent. Again, there is right/left asymmetry of the potential in the last second. An additional potential change occurs in the last 300 msec before activity. Lowest trace: Double standard error of the average brain potentials. Normalization time span –3 to –2 sec as indicated (see methods). Recording vs. linked ears. n = 250 each (VOK 24).

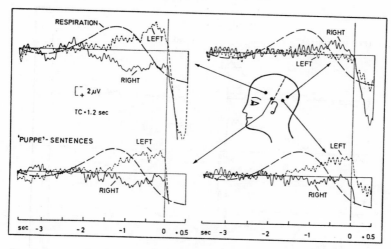

Fig. 6. Topography of averaged brain potentials preceding speech (different sentences). Averaged respiration course is shown by broken lines. The time course difference
between the 2 sides (BPs and R-waves) declines in the anterior-posterior direction. Note
the disappearance of any sign of activity in the most posterior electrode on the left hemisphere. Vertical line indicates onset of activity. Sweep time 4 sec, n = 205. Reference linked
ears (VOK 83).

In addition to the R-wave, the galvanic skin response (GSR) must also
be considered as a possible source of contamination of the cranial potentials.
The R-wave definitely cannot be accounted for by the GSR [GRÖZINGER
et al., 1974a, 1975] and in all but one of our subjects there is less correlation
between GSR and respiration than between brain potentials and respiration.
Furthermore, when the data of a given experiment are analyzed in consecutive subgroups, there is again a greater correlation between brain potentials
and respiration than between GSR and respiration. However, there is some
correlation between GSR and respiration so that the GSR recorded from
the nose or chin sometimes shows similarities to the average respiration
pattern.

Figure 11 illustrates how important routine control of galvanic skin
response is in the search for uncontaminated cerebral potentials. The GSR
is often large at mid-line sites of the head such as the vertex, forehead, nose
or chin. We suspect that some of the slow pre-speech potentials presented at
this symposium and interpreted as glossokinetic potentials and facial
muscle activity [SZIRTES and VAUGHAN, this volume] are in fact GSR.

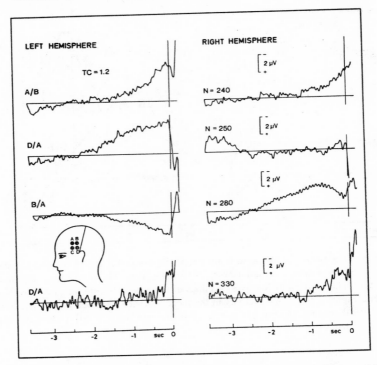

Fig. 7. Four examples of averaged cerebral potentials (Bereitschaftspotential) preceding different speech utterances (upper and second trace 'a', third trace 'Puppe') or articulatory movements (lowest trace) recorded bipolarly from close lying electrode positions over the precentral area, Broca's area and frontal cortex of both hemispheres. The figure documents the occurrence of Bereitschaftspotential differences between closely placed electrodes of the same hemisphere and different potentials over both hemispheres. The third trace shows the R-wave superimposed. Vertical line = activity onset in mouth EMG (VOK 49, 47, 70 and 52).

Glossokinetic potentials certainly do not account for the pre-EMG potentials recorded from the scalp in our experiments since our subjects were trained to avoid tongue movements prior to speech and to begin as sharply as possible. Recording from several facial electrodes allowed exclusion of tongue movements prior to orbicularis oris muscle activity. On the other hand, control experiments involving sudden tongue movements without articulation (fig. 12) showed scalp potentials which preceded the electromyographical activity of the tongue muscles and which must have a cerebral origin.

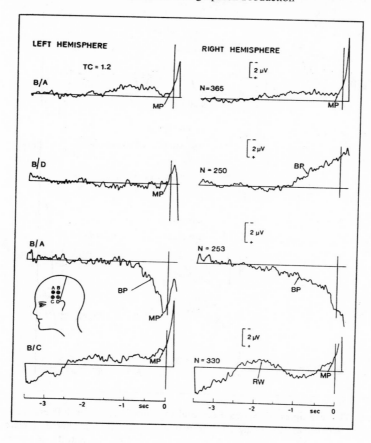

Fig. 8. Four examples of averaged cerebral potentials (motor potential = MP; Bereitschaftspotential = BP) preceding different words or tones (upper trace 'a', second trace to fourth trace 'Lotte') recorded bipolarly from closely placed electrodes over the precentral area representing facial muscles, Broca's area and nearby frontal sites of both hemispheres. The traces show MP differences between closely lying electrodes over the same hemisphere and different MPs and BPs over both hemispheres. Superimposed R-waves are present in the lowest and perhaps also in the uppermost trace. Vertical line = activity onset in mouth EMG (VOK 13, 23, 35, 15).

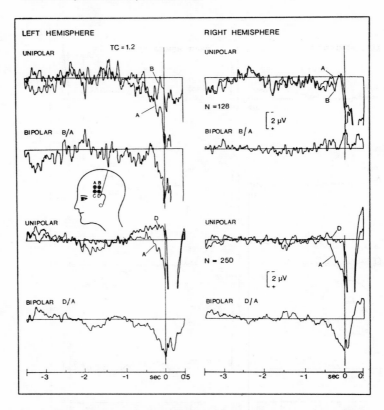

Fig. 9. Two examples of averaged cerebral potentials preceding different words or tones (upper two traces 'humming', lower two traces 'mouth movements') recorded unipolarly vs. linked ears and bipolarly from electrodes over the precentral area representing facial muscles, Broca's area or nearby frontal sites of both hemispheres. The potentials with positive trend in the last half-second preceding mouth EMG onset may represent an analogue to the premotion positivity (PMP) seen prior to finger or eye movements. Vertical line = activity onset in mouth EMG (VOK 51 and 44).

3. Potentials similar to Pre-Motor Positivity and Motor Potentials

In the cerebral potentials preceding finger movement, DEECKE, *et al.* [1969] distinguished three potentials with different distribution and/or onset times: The BP (readiness potential) starting on the average 850 msec prior to movement, the pre-motion positivity (PMP) similarly bilateral and

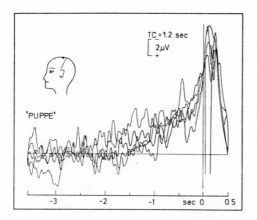

Fig. 10. Constant shape of the BP in five successive subgroups of one experiment with 150 repetitions of the word 'Puppe'. The number in each subgroup is 30. Unipolar recording from the vertex versus linked ears. The vertical line indicates the onset of activity in mouth EMG (VOK 98).

widespread over the hemispheres starting on average about 90 msec prior to movement, and the motor potential (MP) localized over the contralateral precentral hand area starting on average 50–60 msec prior to movement. While no MP has been found preceding eye movements (which corresponds nicely to the totally different motor organization of eye movements as compared to finger movements [KORNHUBER 1974a, b]), a PMP has been found before eye movements although with an earlier onset [BECKER *et al.,* 1972], on the average 150 msec prior to movement. It is sometimes difficult to say whether a positive potential preceding speech is a PMP or a positive BP. However, judging from the latencies, potentials similar to the PMP have been found preceding speech production and vocalization in our experiments (fig. 9). Analysis for stationarity in different subgroups of the experiments, numerous control recordings of extracerebral potentials and bipolar recordings from close electrode positions of the same hemisphere lead to the conclusion that these potentials are of cerebral origin. The onset time of these potentials seems to be even earlier than those preceding eye movements.

With respect to the motor potential, special care must be applied to recordings preceding speech production because of the myogenic potentials of numerous muscles in the head involved in speech and phonation (fig. 1).

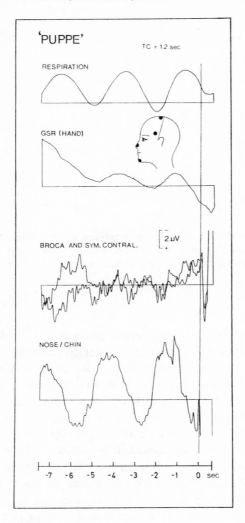

Fig. 11. Correlation of the GSR (recorded from the hand and from nose versus chin) with respiration. With a shorter analysis time, the mid-face GSR would be mistaken as a glossokinetic potential. In this case, the scalp electrode over Broca's area and its contralateral homologue appears to show a BP of about 1.5 sec duration. This trace must be interpreted with care, however, because of the possible contamination with GSRs (n = 40). The vertical line indicates onset of mouth myogram (word 'Puppe'). All curves are filtered by a low-pass 10-Hz filter (VOK K9).

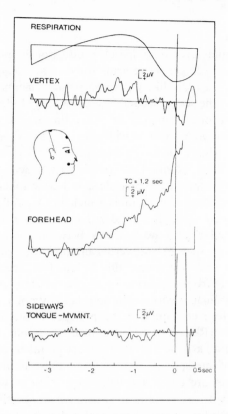

Fig. 12. Control experiment with lateral tongue movements shows the relation of the glossokinetic potential (lowest trace) to average respiration and scalp recordings. The glossokinetic potentials certainly do not account for the pre-activity potentials recorded from the scalp. Note the correlation between respiration and average vertex potentials (n = 40; VOK K6).

However, an extracerebral origin is improbable in experiments with bipolar recordings from nearby positions on the same hemisphere, and with control of galvanic skin response, respiratory potentials and glossokinetic potentials. Such bipolar recordings in good experiments have verified that potentials occur with a sharp change to negativity over the precentral area (fig. 8) where lip-movements and vocalization can be obtained by electrical stimulation of the cortex. However, contrary to the motor potential preceding finger movements, the parallel potential preceding speech may occur bilaterally or unilaterally.

Discussion

The study of cerebral potentials preceding speech production is more complicated than of those preceding finger movements for several reasons. First, the movements of the face are more likely than finger movements to give rise to myogenic potentials of the scalp. Second, there are many hidden movement components such as from the soft palate or the throat; the air burst of the phonation (which has often been considered to be the speech onset and therefore been used to trigger the analysis of the cerebral potentials) starts actually much later than the onset of speech muscle activity. The literature of cerebral potentials supposed to precede speech now appears to have dealt largely with sensory evoked potentials which in fact follow the onset of speech and which were wrongly considered to be pre-speech potentials. For example, the similarity in wave forms between cerebral potentials (interpreted as preceding speech) and verbal utterances or phonemes [SCHAFER, 1967; SZIRTES and VAUGHAN, this volume] is due to sensory evoked potentials resulting from triggering by a voice microphone. Third, the onset of speech is coordinated with expiration, i.e. any potential related to respiration will therefore contaminate the pre-speech cerebral potentials. Respiratory potentials (R-waves) recorded from the scalp are largely of cerebral origin [GRÖZINGER *et al.*, 1975] and must be taken into account in any interpretation of pre-speech cerebral potentials. Fourth, galvanic skin responses sometimes are correlated with respiration as well and, therefore, partly correlated with speech onset. The numerous possible artefacts and potentials of extracerebral origin demonstrate that rigid control of artefacts with multiple channel recordings from the eyes, tongue, midline face, respiration and GSR is essential for recognition of the cerebral potentials preceding speech.

Another important point is a long sweep time for recognition of respiratory potentials and galvanic skin responses. Furthermore, control of stationarity by subgroup analysis of the experiments is necessary. Subjects who cannot maintain steady eye fixation or tongue position and subjects with pronounced galvanic skin responses are not suitable for these experiments.

A considerable amount of data is necessary to reach safe conclusions, even more so than in the case of finger or eye movements. Judging from our experience, all studies with pre-speech cerebral potentials published to date were based on too few samples. Cerebral potentials preceding speech production are undoubtly more complex and variable than potentials preceding a fast, stereotyped finger or eye movement [KORNHUBER and

Deecke, 1965; Deecke, *et al.,* 1969, 1973, 1976; Becker *et al.,* 1972; Deecke and Kornhuber, 1977]. Considering the complex nature of speech, this is not surprising.

Furthermore, it is especially difficult to find an indifferent reference for unipolar recordings. Linked ears are not an ideal reference since they pick up potentials from temporal and basal sites of the hemispheres. Extracephalic references, on the other hand (e.g. vertebra prominens versus sternum, balanced to minimize electrocardiogram), are strongly influenced by galvanic skin potentials and thoracic movements.

One fact emerges clearly from the bewildering picture: The asymmetry of the cerebral potentials preceding the onset of speech starts earlier than the hemispheric asymmetries found before finger and eye movements. While the main asymmetry preceding finger movement occurs in the last 100 msec or so before movement onset, the cerebral potentials prior to speech production are often clearly asymmetrical already 2 sec or more before speech onset. As yet, however, we cannot predict hemispheric dominance for speech from our results. In this connection, it should be kept in mind that it is very difficult to control the cerebral activity of a human subject over a period of several hours as is required for these experiments. The possible continual interplay between verbal and other simultaneous, unassociated mental activity may also influence the recordings.

Dr. B. Grözinger, Section of Neurophysiology, University of Ulm, Oberer Eselsberg, *D–7900 Ulm* (FRG). Tel. (0731) 176 23 34.

Language and Hemispheric Specialization in Man: Cerebral ERPs.
Prog. clin. Neurophysiol., vol. 3, Ed. J.E. DESMEDT, pp. 104–111 (Karger, Basel 1977)

Scalp-Recorded Slow Potential Asymmetries Preceding Speech in Man

MORTON D. LOW and MORGAN FOX

Division of Neurology, Department of Medicine, University of British Columbia and Department of Electroencephalography, Vancouver General Hospital, Vancouver

The language system in the brain of man is presumed to reside primarily in the left hemisphere. This assumption is based upon indirect evidence derived from studies of the effects of cerebral lesions on comprehension and production of elements of speech. There is as yet very little direct electrophysiological evidence of cerebral activity specifically related either to the processing of linguistic input on the receptive side, or to the production of language on the motor side. Several investigators have reported asymmetries of evoked potential configuration between the left and right hemisphere [MORRELL and SALAMY, 1971; BUCHSBAUM and FEDIO, 1969; DAVIS and WADA, 1974] suggesting differences in function of the two hemispheres during perception but without necessarily relating these differences to language mechanisms. However, WOOD et al. [1971] recorded auditory evoked potentials during speech perception from 10 right-handed subjects and concluded that different neural events occur in the left hemisphere during analysis of linguistic versus non-linguistic parameters of the same acoustic signal. TEUBER and MORRELL [1973] have shown that click-evoked potentials are attenuated asymmetrically (more on the left side than on the right) when the subject is listening to sensible speech. These few studies have provided some experimental support for the idea that the left and right hemispheres function differently in performing those linguistic processes necessary for speech perception.

The experimental-physiological evidence for differential hemispheric function during speech production is equally scanty. MCADAM and WHITTAKER [1971] suggested that motor speech function might be localized to one hemisphere, on the basis of asymmetries in the Bereitschafts-potential (BSP) or motor potential preceding the spontaneous articulation of polysyllabic words. Their conclusions have been criticized [cf. SZIRTES and VAUGHAN, this volume] because of the great difficulty of excluding non-cerebral artifacts from recordings of motor potentials occurring during or immediately prior to articulation. GRÖZINGER *et al.* [1975] have shown asymmetries of scalp-recorded slow potentials preceding speech which they believe to be related to central control of respiration prior to and during articulation, but these asymmetries were not systematically related to cerebral dominance (as indicated by functional handedness) [GRÖZINGER *et al.*, this volume]. ZIMMERMANN and KNOTT [1974] have also demonstrated asymmetries in slow potentials recorded from the scalp during and prior to speech, and they have shown that these asymmetries can be different in stutterers as compared to normal individuals.

With the exception of DAVIS and WADA [1974; this volume], all of these studies have used only functional handedness as an indicator of cerebral dominance for speech. While it is generally agreed that left hemisphere structures are mainly responsible for speech in most, if not all, right-handed individuals, the relationship of handedness to language dominance is much less clear in left-handers. As well, the concept of participation of the right (non-dominant) hemisphere in the mechanisms of language is almost totally unexplored in spite of clinical evidence that such participation may be important [KINSBOURNE, 1971].

McADAM and WHITTAKER's observation of asymmetrical BSPs preceding articulation suggested to us that other cerebral slow potentials (such as the CNV) might serve as dynamic indicators of hemispheric participation in language production in the human brain, and because of the much longer duration of the CNV-like potentials as compared to motor potentials, such recordings might also offer a means for avoiding artifacts introduced by preparatory changes in tongue and throat muscle tone.

Because of our previous research experience with cerebral slow potentials such as the CNV, and the clinical necessity for evaluation of certain epileptic patients by the carotid amytal speech localization procedure, we were able to design a series of electrographic studies of cerebral speech dominance using carotid amytal testing as a final control in some subjects.

Methods

Experiments have been carried out to date with 58 subjects ranging in age from 5 to 30 years. The subject population included 43 healthy volunteers selected as 20 'pure'

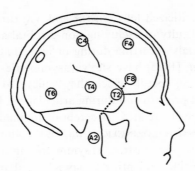

Fig. 1. Scalp location of recording electrodes. T_1 and T_2 are ½ distance from tragus to F_7 and F_8.

left-handers and 23 'pure' right-handers according to the criteria of ANNETT [1967], and 15 epileptic patients admitted to the Vancouver General Hospital for evaluation for surgical treatment of medically intractable seizures. Of the 15 patients, 11 were right-handed and 4 left-handed. Volunteer subjects were recruited from the University community (Faculty, Staff and Students) and each was paid a nominal honorarium for participation in the study. The group of normal subjects included two sets of identical twin pairs. These were 8-year-old male children. One of the pairs was mirror-image with one of the siblings right- and the other left-handed. One of the other boys was right-handed while his twin sibling was ambidextrous. An older brother of this pair was functionally left-handed and was also included in the study.

Each subject was run in at least two recording sessions. During each session, scalp EEG activity was recorded from International 10–20 system electrode positions F_7, F_8, T_3, T_4, CZ and from positions which we have designated T_1 and T_2 (fig. 1) using a contralateral ear reference and gold-plated silver electrodes. The amplifier system (Beckman Type R) time-constant was 4 sec. During the test series, the subject sat in a comfortable chair in a sound-deadened room facing a translucent screen with his eyes open. The subjects were instructed to visually fixate a small target in the center of the translucent screen. The EOG and the glossokinetic potential were recorded and averaged simultaneously with the scalp activity. All averaging was done off-line from tape recorded data using a Fabritek 4-input signal averager. Averages were written out with an X-Y Plotter for comparison and measurement.

Each test session lasted about 30 min and included 6 paradigms. The paradigm of interest was the standard cued-reaction time $S_1 \rightarrow S_2 - R$, where S_1 was a single stroboscopic light flash (Grass PS2, intensity setting 8), S_2 was a one-second tone burst at 1,500 Hz, and approximately 70 dB intensity, the $S_1 \rightarrow S_2$ interval was 1.4 sec and R was an instructed word. Paradigm 5 consisted of 33 trials of $S_1 \rightarrow S_2$ with the subject responding to the tone by saying the word 'hi'. Paradigm 6 was the same with the subject responding to S_2 by saying his first name. These trial sets were preceded by a series of 60 flashes alone, 60 tones alone, 33 presentations of flash followed by tone and 33 $S_1 \rightarrow S_2$ pairs with the subject responding to S_2 with a button-press.

Averaging was done in blocks of 32 trials with a 5-sec epoch beginning 0.5 sec prior to the onset of S_1. X-Y plots of averaged traces from T_1, F_7, and T_3 electrode positions were compared to traces from T_2, F_8, and T_4 positions using area measurements as the units for comparison [Low, 1969]. No measurements were made of scalp recorded activity if the EOG or GKP channels showed more than 5 microvolts (peak) baseline deflection during the $S_1 \rightarrow S_2$ interval. Following EEG testing, the epileptic patients eventually were subjected to bilateral carotid amytal tests for language dominance. Determination of dominance with carotid amytal tests was made without the knowledge of the results of electrographic recording.

Results

Epileptic patients. Two *a priori* assumptions were made in assessing the electrographic data from epileptic patients. The first was that the intra-carotid amytal tests would provide conclusive evidence regarding cerebral dominance for language production. The second assumption was the averaged EPs would be asymmetrical, showing greater negative shifts on the dominant side (fig. 2). On the basis of this assumption, the electrographic finding correlated exactly with the conclusion of carotid amytal tests in 12 out of 15 patients (table I). The three errors were all prediction of right dominance in right-handed patients shown ultimately to be left-dominant.

Volunteer subjects. The areas of averaged negative shifts occurring between S_1 and S_2 (for above-baseline negativity up to 1.2 sec following S_1), were measured for electrode positions T_1 and T_2. The results (overall mean values including measures from both runs for all subjects) were as shown in tables II and III. The T_1 and T_2 means differed significantly at the $p = 0.05$ level.

While there was a significant difference in grouped temporal CNV measurements which was in the expected direction, i.e. greater negativity on the side opposite the subject's handedness both in children and adults, not all individual recordings showed this relationship. Within subjects, there was considerable variability both in direction and degree of asymmetry of slow potential magnitude between left and right hemispheres.

In determining language dominance, we took the most consistent left-right relationship over the 4 speech-response paradigms for any given individual. More consistent asymmetries were detected during the second recording session than during the first, and during the paradigm requiring a response with the subject's given name than during the paradigm requiring the more automatic greeting response.

The recordings from the identical, mirror-image twin pair showed more consistent negativity on the left side than on the right in the right-handed child and more consistent negativity on the right in the left-handed sibling.

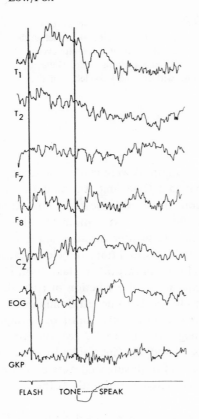

Fig. 2. Sample of averaged trials from an 8-year-old normal child showing negative shift in the S_1-S_2 interval at T_1, while simultaneously recorded and averaged traces from other electrode locations show no significant S_1-S_2 negativity. Flash-tone interval = 1.4 sec; height of tone code = 8μV. Scalp electrodes referred to contralateral ear.

The recordings from the other twin pair (one right-handed, the other ambidextrous) showed consistent mixed asymmetries with greater negativity on the left preceding automatic speech and greater negativity on the right preceding articulation of their Christian names.

Discussion

Inspection of the data from more than 100 recording sessions has led us to conclude that while there appears to be a slow potential recordable from the posterior-inferior frontal regions prior to speech in a manner analogous to the vertex CNV in the more usual cued-reaction time paradigm, and while these 'speech CNVs' do show asymmetries which can be related to cerebral dominance for language production, the procedures involved

Table I. Electrographic vs. carotid amytal determination of hemisphere dominance for speech (15 patients)

CNV test prediction		Carotid amytal test results
Right-handed Left-dominant	} 7	$7 + 2^1$ confirmed
Left-handed Right-dominant	} 2^1	1
Right-handed Right-dominant	} 4^1	2
Left-handed Left-dominant	} 2	$2 + 1^1$

[1] CNV test error.

Table II. Comparison of mean speech-CNV areas recorded from T_1 and T_2 electrode positions in 11 right-handed and 11 left-handed normal adult subjects aged 18–30 years. Mean values include measures from two recording sessions for each subject. Units of area are result of integration of negative amplitude as a function of time over 1.2 sec following S_1

Handed	Scalp position	Mean 'CNV' area
Right (n = 11)	T_1	12.3^1 units
	T_2	7.5
Left (n = 11)	T_1	7.0
	T_2	15.3

[1] T_1 vs. T_2 means significant difference at $p = 0.05$.

Table III. Comparison of mean speech-CNV areas recorded from T_1 and T_2 electrode positions in 12 right-handed and 9 left-handed normal children aged 5–14 years. Mean values include measures from two recording sessions for each subject. Units of area are result of integration of negative amplitude as a function of time over 1.2 sec following S_1

Handed	Scalp position	Mean 'CNV' area
Right (n = 12)	T_1	10.5^1 units
	T_2	5.5
Left (n = 9)	T_1	5.5
	T_2	14.0

[1] T_1 vs. T_2 means significant difference at $p = 0.05$.

can present serious methodological difficulties. In our experience, there is a high probability of contamination of the scalp-recorded EEG activity (from the locations used in this study) by horizontal eye movement artifact. Unlike the standard CNV recording situation, vertical eye movements contribute almost nothing to the T_1, T_2, F_7, and F_8 channels. On the other hand, we found that glossokinetic potentials were very rarely asymmetrical, and were in fact rarely present in the time interval during which our measurements were made. KNOTT and ZIMMERMANN [1974] have recently shown (using cine-radiography) that the earliest detectable movement of any of the muscles of phonation is that of the palate, and this occurred in their study approximately 200–400 msec prior to articulation. Our CNV measurements excluded all or most of this immediate pre-articulation interval.

We have as yet no satisfactory explanation for the errors in prediction of hemisphere dominance in the three epileptic patients. All three had very abnormal resting EEGs, and the abnormalities included frequent focal discharges in the right temporal-frontal area. It is possible that our slow potential recordings were contaminated by the ongoing asymmetrical abnormal activity. With our present technology, we see no ready solution to the problem of this kind of contamination.

In many of the individual recordings, coincident with the negative going shift in the $S_1 \rightarrow S_2$ interval on the 'dominant' side, there was a positive going shift on the 'non-dominant' side. This observation has an interesting parallel in the cerebral blood flow studies of INGVAR [1974]. He has shown that cerebral blood flow increases from resting levels in certain regions of the dominant hemisphere during reading and speaking, and coincidently blood flow in homologous regions of the non-dominant hemisphere are decreased from resting levels. One could infer that both the blood flow changes which he documents and the electrical changes which we observed are reflecting alterations in regional neuronal function associated with the processing and generation of language.

Of course, the specific origin and functional significance of the slow potentials which we are recording from inferior-posterior frontal regions preceding speech are still open questions. We are sufficiently concerned about artifact contamination and reliability of dominance prediction using such recordings that we have completely automated our data acquisition and analysis procedures, and have adopted a new paradigm for elicitation of the 'speech CNV'. The entire experiment is computer controlled (PDP 11/20) and includes on-line digitization and storage of EEG, EOG, and GKP signals with automatic rejection of trials which contain more than a predetermined baseline excursion in the EOG or GKP leads. In the new paradigm, S_1 is a word presented on a video screen for 40 msec. 1.5 sec later, a brief tone pip (S_2) occurs and the subject is required to speak the word he has just read. The $S_1 \rightarrow S_2$ series includes 15 different words equated for 'motivation loading' and the system therefore requires constant visual attention to the stimulus source as well as reading and motor speech. This new paradigm has now been used in 70 subjects, including 4 epileptic patients, 30 normal adults and children, and 36 other children with mildly abnormal neurological findings. Results were similar to those reported above for the original procedure. There was one error in predicting dominance in the 4 epileptic patients.

The fact that asymmetries in these slow potentials can be demonstrated in children as young as 5 years of age, even in the absence of a vertex CNV, agrees well with the previously published observations of Low and STOILEN [1973] and with the clinical observation that cerebral dominance for language production appears to be established by about the middle of the first decade of life.

If these slow potentials do reflect neuronal activity which underlies mechanisms of language production, our observations would support the idea that both left and right hemispheres contribute to the language system in a dynamic way and perhaps in different ways from moment to moment. Our experiments are continuing, with particular emphasis on determination of the topographic distribution of these slow potentials and their relative configuration in the presence of brain lesions both congenital and acquired.

Acknowledgments

The authors gratefully acknowledge the assistance of JANICE GALLOWAY, SIEW YONG TH'NG, MICHAEL BAKER, and FRANCES BURTON, and the professional cooperation of Dr. J.A.WADA. This work was supported by Grant MA-3313 from the Medical Research Council of Canada.

Dr. M.D. Low, Department of Electroencephalography, Vancouver General Hospital, *Vancouver, B.C. V5Z 1M9* (Canada). Tel. (604) 876 32 11.

Language and Hemispheric Specialization in Man: Cerebral ERPs.
Prog. clin. Neurophysiol., vol. 3, Ed. J. E. DESMEDT, pp. 112–126 (Karger, Basel 1977)

Characteristics of Cranial and Facial Potentials Associated with Speech Production

JOZSEF SZIRTES and HERBERT G. VAUGHAN, jr.

Departments of Neurosciences and of Neurology, Albert Einstein College of
Medicine, Bronx, N.Y., and Institute of Psychology of the Hungarian Academy of
Science, Budapest

Several investigators have ascribed averaged scalp recorded potentials
associated with speech production to neural activity arising in part from
cortical areas implicated in the programming of speech. ERTL and SCHAFER
[1967] described phasic potentials that preceded verbal utterances by as
much as 300 msec and varied in waveform with different speech sounds
[SCHAFER, 1967]. These authors later retracted their attribution of these
speech-related potentials to a cerebral source when they observed potentials
of similar timing and waveshape from recordings taken near the mouth
[ERTL and SCHAFER, 1967].

McADAM and WHITAKER [1971a] focussed upon the slow potential shifts
preceding speech, which they considered analogous to the initial negative
shift of the motor potential [KORNHUBER and DEECKE, 1965; GILDEN et al.,
1966]. Mean data from five right-handed subjects showed a significant
asymmetry in the pre-speech potentials, with a relative negativity at electrode
sites overlying Broca's area and the left precentral motor cortex with
respect to the homologous non-dominant sites. Low and Fox [this volume]
and ZIMMERMANN and KNOTT [1974] also reported consistent asymmetries
in slow potential shifts preceding speech in a CNV paradigm.

McADAM and WHITAKER's study was criticized by MORRELL and
HUNTINGTON [1971] for failure to consider the possibility that extracerebral
potentials might have contaminated the pre-speech activity they measured.
An even stronger attack on these findings was made by GRABOW and
ELLIOTT [1974] who considered the scalp recorded activity to be 'glossokinetic potentials' [KLASS and BICKFORD, 1960]. These authors were unable
to confirm the presence of pre-speech asymmetries, attributing this reported
finding to possible lateral deviations in the tongue movements accompany-

ing speech. MORRELL and HUNTINGTON [1972] took a position apparently at variance with their earlier criticism of McADAM and WHITAKER. In their later report, they asserted that the phasic potentials associated with speech have a timing and distribution more compatible with a cortical than extra-cerebral origin. They presented no quantitative data on hemispheral asym-metries, commenting only that such differences were not marked but when present the potentials were most commonly larger over the left hemisphere. These authors apparently referred to asymmetries involving either the pre-speech shifts or the later phasic potentials.

The present study was originally undertaken with the intent of defining the cortical sources of speech-related potentials by utilizing the topographic techniques applied by one of us [VAUGHAN, 1974] to the analysis of various event-related cerebral potentials. Earlier experience with potentials asso-ciated with contractions of cranial musculature [VAUGHAN et al., 1968] had alerted us to the problem of possible artifactual contamination by extra-cerebral potentials, so special efforts were made both to control and monitor activity which might arise from the structures involved in the speech act. As the study progressed it became apparent that the distribution and timing of extracranial potentials were critical issues in the interpretation of speech-related activity recorded from electrodes overlying the cortical regions thought to be implicated in speech production. This report outlines our experience with speech-related potentials recorded from cranial and facial sites.

Methods

A total of 18 subjects were studied, 11 Hungarian and 7 American. 15 subjects were right-handed and three left-handed, as ascertained by history and observation of manual preferences. In six subjects, detailed mapping was carried out, utilizing various electrode arrays and reference sites, as described below. In the remaining subjects, a standard bilateral array was employed to evaluate hemispheral asymmetry.

The standard array comprised a set of symmetrical bilateral placements overlying the presumed locations of Broca's and Wernicke's areas of the dominant hemisphere, the inferior portion of the precentral gyri, inferior frontal sites 5 cm anterior to the 'Broca' placements, frontal parasaggital loci 5 cm anterior and lateral to the vertex, and over the zygomatic arch 4 cm anterior to the tragus. The linked ears were employed as reference for interhemispheral comparisons. More detailed mapping studies employed additional place-ments which included electrodes on the nose, cheek and cranial sites arranged so as to permit isopotential mapping of the amplitudes of the various potentials. Comparisons were made among recordings using linked ears, mid-occiput and a sterno-vertebral balanced electrode pair as references. The non-cephalic reference was used for isopotential mapping. Some recordings were done with bipolar linkages to evaluate local potential gradients. In addition to the foregoing recording sites, bipolar surface electrodes were placed to record EMG activity from the larynx, lips, and the base of the tongue. The signals were amplified employing systems with time constants which exceeded 3 sec. High frequency response was down 3 dB at 60 Hz or more for all recordings. Speech-related potentials were extracted from the raw records by averaging with respect to the onset of the speech sounds picked up by a microphone placed close to the mouth. Comparisons were made with recordings triggered from the larynx and from the EMG recordings, but the phonogram was most reliable for synchronizing the speech-related potentials. The relative timing of the sound and muscle activity was assessed by rectifying the phonogram and EMG prior to averaging so as to obtain a true envelope of these signals. All signals were recorded on analog FM tape. 25–100 trials for a designated length of record, usually 1 sec or more preceding and following the speech sound onset, were averaged employing a DEC Linc 8 or Hungarian NTA-512 computer. The records were visually edited prior to averaging and any epoch displaying detectable ocular activity or movement artifacts was eliminated.

Activity was recorded in association with a wide variety of speech and non-speech sounds, as well as potentials synchronous with various lip, tongue, and jaw muscle contractions. The latter recordings utilized the rectified EMG recorded from appropriate sites for synchronizing the averaging process. The distribution of the potentials associated with muscle contraction was evaluated utilizing the same electrode montages as for the speech-related potentials.

Results

Speech-related potentials were recorded from all sites over the cranium and facial region. The morphology and distribution of these potentials varied widely as a function of the specific utterance and showed individual differences as well. Due to the extensive distribution of speech-related activity, the reference site also importantly influenced the characteristics of the potentials. Invariably, the activity time locked to speech sounds began substantially prior to the phonogram, often as much as 1 sec, and also outlasted the acoustic signal by several hundred msec.

Morphology of speech-related potentials. Investigators have distinguished an initial slow potential shift from the later complex of phasic potentials, concurrent with the main features of the speech act. However, the differentiation of slow from phasic components is often arbitrary and the relative timing of these segments of speech-related activity varies widely with different speech sounds. Furthermore, the amplitude and polarity of the initial slow shift varies across subjects for nominally similar utterances, as

well as in the same subject for different utterances. In figure 1, potentials
recorded from the left inferior frontal electrode overlying Broca's area are
depicted for one subject uttering six different speech sounds. Striking
variations in waveshape are evident.

    Although earlier slow shifts are not well developed in this subject
except in association with /ɛt/, the onset of a large phasic component, either
positive or negative in polarity, precedes sound onset by at least 500 msec.
The particular speech sounds whose associated potentials are depicted were
selected to illustrate not only the markedly different potentials associated
with specific speech sounds, but also the fact that phonemic combinations
produce waveshapes which appear to be summations of the potentials
related to the component sounds. Thus, the phasic potentials associated
with /bɛt/ comprise an initial positive wave characteristic of the plosive
utterance /b/, followed by a negativity associated with /ɛ/ and finally a
sharp positive wave representing /t/. It will be noted that the initial positivity
in /bɛt/ is smaller than that in /bi/. This appears to be due to the partial
overlap of the waves associated with the following vowels, the /ɛ/ potential
being a negative deflection and /i/ a positive one.

    In subjects with well-developed initial slow shifts, these potentials were
as much affected in amplitude and polarity by the particular utterance as

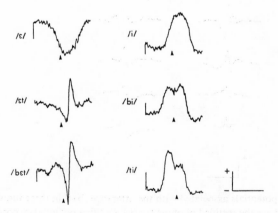

Fig. 1. Speech-related potentials recorded from the left inferior frontal (Broca's area)
placement in association with the utterance of the indicated speech sounds. Non-cephalic
reference. Onset of speech sound is indicated by the arrow beneath each trace. Each record
is the average of 50 trials. Subject J.K. Calibration: 20 μV, 1 sec.

were the later phasic components. The amplitude and polarity of the slow shifts were rather stable for each utterance in a given subject, but varied across subjects for the same speech sound, as seen in figure 2. Despite the intersubject variability of the slow potential shifts, the waveshape of the phasic components for a given speech sound showed a fair degree of similarity across subjects, as shown in figure 2.

Distribution of speech-related potentials. Our initial observations revealed such a wide cranial and facial distribution of speech-related activity that it was immediately apparent that no electrode situated on the head could serve as an indifferent reference. Thus, a non-cephalic reference was employed for evaluating the potential distributions, recognizing that this reference might also be contaminated by widely conducted speech-related potentials. Figure 3 depicts a typical set of potentials recorded from

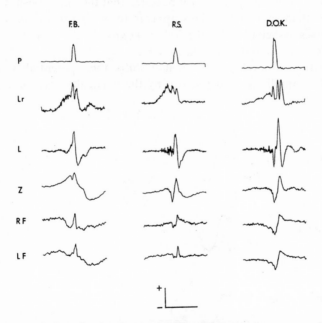

Fig. 2. Speech-related potentials associated with the utterance /bɛt/ in three individual subjects. Signals depicted are the rectified phonogram (P), rectified potentials recorded from the lips (Lr), unrectified lip potentials (L), potentials recorded from the left zygoma (Z), right and left inferior frontal recordings (RF, LF). Z, RF, and LF referred to linked ears. Calibration: Amplitude L, Z, RF, and LF 20 μV except for L and Z subject FB 40 μV. Time: 1 sec.

various sites on the face as well as from the standard inferior frontal placements. The largest potentials are recorded from the nose and the waveforms
recorded from the cheek and the inferior frontal sites have essentially the
same waveform. The activity recorded from the chin shows a similar waveform, but is inverted in polarity. Not only the phasic negativity, but also
the slow negative shift which begins almost 1 sec before speech onset are
mirrored in the positivity recorded below the oral cavity. These observations
provide evidence for a major source of speech-related potentials in structures
situated between the nose and chin. Reference to figures 2 and 3 indicates,
however, that the waveforms recorded from the face and cranium are not
identical and must, therefore, reflect potentials derived from more than
one source. This inference is strengthened by observations of the distribution
of potentials associated with different speech sounds. The positive wave

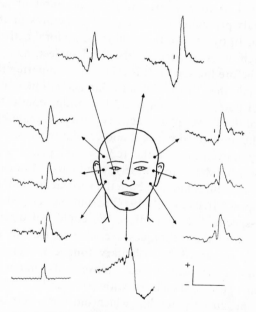

Fig. 3. Potentials recorded from the indicated sites associated with the utterance /ɛt/.
Non-cephalic reference. Average of 50 trials synchronized with speech sound onset indicated by phonogram. Onset of sound indicated by a vertical line for each trace. Note the
remarkable similarity in waveform of the potentials recorded from the nose, right cheek
and right inferior frontal placements. A slow negative potential shift precedes sound by
some 700 msec in the upper facial and cranial leads. A positive shift is present over the
same interval in the chin recording. Subject J.K.

recorded with /i/ sounds is also larger over the lower face than the cranial sites, but does not invert in polarity below the mouth. Thus, the configuration of the generators of the /i/ potentials must differ from those of the /ɛ/ potentials, although the potentials are largest over the lower face in both cases.

   Relationship of speech potentials to muscular activity. Highly complex sequences of muscular activity produce changes in the configuration of various portions of the vocal tract accompanying phonation and articulation. These muscular activities involve the respiratory apparatus, the larynx, pharynx, tongue, and mandible. Our study sampled only a small portion of potentials associated with activation of the lips, jaw, and tongue so that a full picture of possible movement artifacts was not obtained.

   In figure 2, activity recorded from the perioral region is depicted both as directly averaged potentials and as an average of rectified potentials. The latter technique permitted us to estimate precisely the onset of any muscular activity or potentials preceding movements of structures in the anterior region of the mouth. In two of the three subjects, perioral activity begins 600 msec before speech onset, whereas in the third subject, activity can be seen to increase 1 sec before the sound. It is evident in comparing the rectified and unrectified signals that this early muscular activation is not associated with slow perioral potentials which could be volume-conducted to more distant sites. Some of the slower phasic activity, however, manifests similar waveforms in both the unrectified perioral and cranial records.

   Potentials generated by movements of the tongue and palate were considered the most likely candidates for widely conducted slow and phasic potentials, since these structures move in both the vertical and antero-posterior directions during speech [KENT *et al.*, 1974] permitting the generation of complex movement-specific field potentials. We could not directly evaluate palatal movements with the techniques available to us, but did compare averaged potentials generated by voluntary tongue movements with speech-related potentials. Vertical tongue movements mimicked closely the field depicted in figure 3 associated with /ɛ/ sounds, whereas anteroposterior movements produced potentials which did not invert over the anterior face, as in the case of /i/ sounds. Voluntary deviation of tongue movement to either side produced marked asymmetries in the facial and cranial distribution of the associated potentials. Thus, the form and distribution of facial and cranial potentials associated with speech production are compatible with a partial origin in the tongue movements which precede and accompany speech production.

Contraction of the jaw musculature and consequent jaw movements represent another possible source of scalp potential contamination during speech, especially in view of the position of the temporalis muscles in relation to Broca's area. Isometric contraction of these muscles produced by clenching the closed jaw was associated with large phasic potentials which were recorded over the entire face and cranium. The distribution of this activity differed somewhat from that associated with tongue movements, being maximum over the posterior frontal areas and lateral facial regions overlying the musculature, in contrast to the tongue potentials which were largest in the midline of the face. Asymmetries in the amplitude of potentials associated with jaw clenching were frequently observed.

Asymmetry of speech-related potentials. Asymmetries were frequently observed in both antecedent slow shifts and phasic components at homologous cranial and facial sites, as seen in figures 2 and 3. Figure 4 presents an isopotential amplitude plot of an initial negative shift manifesting asymmetry over the frontal region. Although the maximum amplitude overlies the midfacial region, the potentials recorded over the left anterior hemicranium are larger than over the right. In this instance, but not invariably, similar asymmetries in distribution were found for the later phasic deflections which accompanied speech.

Measurements of the pre-speech slow potential shifts at the inferior frontal and zygomatic sites in 10 right-handed subjects provided the results

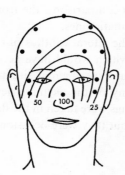

Fig. 4. Isopotential amplitude distribution of the pre-speech negative shift associated with the utterance /ε/ in one subject. Amplitudes are compared to the largest potential which was recorded from the nose. Smooth isopotential curves have been drawn at 75, 50, 37, and 25% of the maximum amplitude interpolated from measurements of records taken from the indicated electrode sites. Note the asymmetrical extension of the field over the left frontal region.

Table I. Amplitudes in μV of pre-speech slow shifts recorded from inferior frontal and zygomatic electrodes

Subject	Inferior frontal		Zygomatic	
	left	right	left	right
1	2.5	7.5	7.5	14.5
2	4.0	2.3	7.6	6.3
3	5.1	0.3	9.0	3.5
4	2.7	4.4	4.4	2.2
5	0.2	−2.0	8.0	8.5
6	−3.0	2.2	−0.8	7.3
7	6.0	6.0	6.2	6.2
8	−1.5	−1.5	1.5	8.5
9	−0.3	0.9	−1.0	1.8
10	−2.9	−4.7	8.5	4.1

presented in table I. There was no intersubject consistency in polarity of the slow shifts nor of amplitude asymmetry. In four subjects, the right frontal records were relatively negative with respect to the left, in four the reverse was the case, and in two there was no asymmetry. In seven subjects, the relative amplitudes recorded at the zygoma were in the same direction as in the frontal records, whereas in three subjects the asymmetries were discordant. Usually, the zygomatic potentials were larger than the frontal ones, but there were some exceptions. In 18 of 20 comparisons, the frontal shifts were relatively more negative than the zygomatic ones.

Recordings taken from electrodes placed at the lateral canthus of each eye failed to show pre-speech polarity reversals which would have accompanied lateral deviations of the eyes. Thus, in our experimental situation, lateralized ocular shifts could not account for any of the asymmetries in the inferior frontal recordings. Even when recordings were taken in the dark, we did not observe pre-speech ocular deviations, but they were occasionally seen following speech.

Discussion

The results reported here lend strong support to suggestions that scalp-recorded speech-related potentials either represent activity of solely extra-cranial origin or are heavily contaminated by such activity. Three lines of evidence have been adduced to reach this conclusion: (1) the occurrence of substantial changes in speech-related potential morphology with different utterances; (2) the distribution of speech-related potentials with maxima overlying the lower face and polarity inversions across the oral region for some speech sounds, and (3) the similarities in form and distribution between speech-related potentials and non-vocal movements of the speech musculature.

Speech-related potentials have been compared to the cortical potentials associated with other voluntary movements. GRÖZINGER et al. [1974] asserted that 'all typical cortical potentials preceding voluntary limb movements could be reproduced'. These authors reported an asymmetry in the readiness potential preceding vocalization, corresponding to hemispheric dominance. They also noted very slow asymmetrical potentials correlated with respiratory movements preceding speech. In subsequent studies, GRÖZINGER et al. reported a greater complexity of pre-speech potentials and a rather variable lateralization which was apparently affected in some way by interactions with the respiratory 'R-waves' [GRÖZINGER et al., this volume].

Since speech requires the activation of musculature sharing a common cortico-bulbar mechanism with non-speech movements of the relevant musculature, it is to be expected that cortical motor potentials would, in fact, be associated with both speech and non-speech muscular activation. VAUGHAN et al. [1968] recorded potentials associated with isometric tongue contractions and with lower facial movements, but noted the importance and difficulty of excluding contamination by extracranial potentials. Indeed, any gross movement of the tongue was found to produce contamination of the cranial recordings. The main evidence for a cortical origin of the potentials associated with tongue and face contractions was the presence of the characteristic motor potential waveshape and a topographic distribution compatible with an origin in the lower portion of the precentral gyrus. In marked contrast to these observations, the morphology of the speech-related potentials changes drastically with the nature of the utterance. Rather than the invariably negative initial slow potential shift, the slow activity preceding speech may be either positive, negative, or absent altogether, depending upon the utterance and the individual subject.

The extraordinary variation in form of the phasic components is also in striking contrast to the regular negative-positive sequence of the motor potential. Although increases in contraction duration or successive flexion-extension movements of the extremities increase the complexity of later motor potential activity, the initial negative-positive sequence is unaffected by the nature of the initial movement. Thus, the polarity reversal of the initial phasic potential with different speech sounds is contrary to the experience with cortical motor potentials associated with contractions of the extremities, tongue, and facial musculature. Although it can be argued that the speech-related potentials represent a distinct process concerned with the programming of speech sounds which precedes the cortico-bulbar motor command, it does not seem plausible that such marked changes in waveshape could reflect differences in net cortical neural activity associated with differences in the verbal engrams. It should also be noted that the speech-related potentials obtained by us and other investigators are substantially larger than the motor potentials, the latter rarely exceeding 10 μV in maximum peak-to-peak amplitude. This circumstance also seems implausible for a speech encoding potential, since high amplitude scalp-recorded potentials require an extensive area of synchronous cortical activation, which seems somewhat unlikely, although not inconceivable. In any event, should extra-cranial components represent a significant portion of this phasic activity, the large size and variable morphology of the speech-related potentials would substantially obscure concurrent potentials of cortical origin.

The data on distribution of the speech-related potentials render the conclusion of substantial extracranial contamination of cranial recordings virtually inescapable. Even if one were to accept the unlikely proposition that the extracranial and cortical activity had virtually the same timing and morphology, the existence of gradients such as those depicted in figure 3 and 4 for speech-related potentials of widely varied morphology indicates that volume-conducted spread to the cranium must occur. Although we looked very hard for some evidence of a distinct focus of activity overlying the speech and motor cortical regions, we could not obtain convincing evidence of a secondary source of potentials distinct from those of clearly extracranial origin. It is important to note that not only the phasic components but also the initial slow potential shifts have distributions indicative of at least a partial extracranial origin. Since the studies of speech-related activity which have endeavored to obtain evidence for lateral asymmetries related to cerebral dominance have focussed upon these shifts, it is important to establish the degree to which potential shifts preceding speech might be contaminated by extracranial potentials. Recordings from facial electrodes have clearly established the presence of slow shifts of either polarity which appear as early as any slow shifts recorded from cranial sites (fig. 2, 3). Although the origin of these extracranial shifts is likely to be a complex one in view of the varying distributions and polarities found for different subjects and utterances, it is evident that measurements of baseline shifts during the second prior to emission of a speech sound are likely to be contaminated by slow shifts of extracranial origin.

The final piece of evidence for the extracranial origin of the speech-related potentials is provided by the comparisons of the potentials associated with the non-speech movements and muscular contractions. These observations merely confirmed what every electroencephalographer knows, namely that tongue, eye, and jaw movements can produce gross contamination of recordings taken from cranial electrodes. More important is the demonstration that the morphology and distribution of slow potentials associated with various muscular maneuvers can closely mimic speech-related potentials. Since the actual spatiotemporal movement patterns of articulation are quite complex [FRITZELL, 1969; KENT et al., 1974], no movement of a single structure is sufficient to simulate the movement patterns associated with speech. Indeed, it is the complexity of the latter which probably gives rise to the varied waveforms and craniofacial distributions of the potentials associated with different utterances. Finally, we note that asymmetrical potentials are readily generated by non-speech movements. Although we know of no systematic study of movement asymmetries in normal speech, the very existence of the motor asymmetry associated with handedness suggests that some degree of asymmetry could occur in the contraction of cranial musculature, despite the bilaterality of cortico-bulbar innervation. Indeed, it is not entirely implausible to suppose that systematic asymmetries in muscular activity during speech could be correlated with hemispheral dominance. Thus, consistent lateralization of extracranial shifts preceding speech could conceivably be correlated with handedness, but not directly reflect cortical activity.

Our failure to observe any consistent lateral differences in amplitude of pre-speech slow potential shifts conflicts with reports of a greater negativity over Broca's area than over the homologous non-dominant site [McADAM and WHITAKER, 1971a]. MORRELL and HUNTINGTON [1972] and GRABOW and ELLIOTT [1974] also found no evidence for consistent pre-speech asymmetries. In contrast, Low et al. [1976] and ZIMMERMANN and KNOTT [1974] report a consistent negativity over the dominant inferior frontal region during the interstimulus interval in a 'CNV' paradigm requiring verbalization following S2. As in all of the speech potential studies, these results are based upon a small number of subjects. Furthermore, the number of epochs averaged in ZIMMERMANN and KNOTT's study ranged from 6 to 12, after the remainder of 40 trials had been discarded due to artifactual contamination. Our experience suggests caution in accepting conclusions based upon concordance of results in a small number of subjects from data based upon small numbers of trials. In our initial assessment of hemispheral asymmetry, five right-handed subjects were studied, four of whom demonstrated a relative negativity over the left inferior frontal region in a majority of averaged records. This study was completed prior to our recognition of the seriousness of extracranial contamination, so that we had not obtained recordings from bilateral facial placements. A replication of this study, with the additional ten subjects reported here, failed to confirm the earlier findings. Thus, of 15 right-handed subjects in our studies, 8 showed a preponderant relative negativity over the presumably dominant hemisphere, 4 over the right and 3 showed no asymmetry. Furthermore, variations in relative amplitudes over the two hemispheres occurred within subjects in replicated runs of 50 trials, as well as with different utterances. Thus, we conclude that in the self-initiated verbalization procedure employed in our studies, evidence for consistent hemispheral asymmetry associated with handedness is lacking. It is possible that the CNV paradigm might permit the recording of cortical potentials without contamination by extracranial potentials. In these tasks, an interstimulus interval exceeding 1 sec has been employed and the verbal response follows S2. Unfortunately, however, the studies employing this paradigm have not recorded from facial sites so that preparatory adjustments of vocal tract structures with accompanying slow shifts prior to S2 cannot be excluded. With such control recordings, replication of the reported observations could be accepted as valid evidence for a lateral predominance of cortical activity preceding speech.

All in all, the experiences reported in this paper are discouraging, since our initial purpose was to observe gross potential correlates of speech encoding. Even if the CNV studies should prove valid as indices of hemispheral dominance for speech, this information, while of possible clinical value, leads to no advance in our understanding of cortical mechanisms underlying speech production. Our concluding remarks will evaluate the possibilities for overcoming the difficulties in the recording and interpretation of speech-related cortical activity disclosed by our experience to date.

In retrospect, it seems naive to have expected that cortical potentials associated with speech could be recorded uncontaminated by extracranial activity. We relied too heavily on the possibility of excluding such artifacts by editing of the raw data, ignoring the potency of the averaging technique for detecting far field potentials of small amplitude which cannot be discerned in the direct recordings. Furthermore, efforts to reduce movement artifacts were unavailing since movements are intrinsic to, and necessary for, speech production. Thus, although some movements of the vocal tract can be restricted, they cannot be eliminated without eliminating the vocalization as well. What possibilities remain for the study of cortical potentials associated with speech?

Obviously, the most direct approach will have to be the utilization of the rare opportunities for intracranial recording of speech-related cortical potentials in conscious patients. This enterprise is a justifiable one, since proper identification of the speech areas is one of the principal requirements of cortical surgery which might encroach upon language mechanisms. Thus, not only would such studies provide us with potentially important information on the cortical dynamics of speech mechanisms, but would also be a potentially valuable tool for the neurosurgeon in guiding the extent of cortical excisions. In view of the far field conduction problem, the use of transcortical recording electrodes is mandatory [STOHR and GOLDRING, 1969]. Through the recording from multiple sites in and around Broca's area, motor cortex, and in the posterior speech zones during articulated speech, important data on the timing of and localization of cortical activity related to speech acts could be obtained.

In the absence of direct intracranial observations, further efforts to interpret scalp-recorded data will be severely hampered. About the only possibility which has not as yet been explored in detail is the use of close bipolar recording chains in the hope of selectively eliminating far field potentials and defining local gradients related to zones of underlying cortical activation. This endeavor might be worthwhile if done in a sophisticated manner. Our preliminary experience, however, suggests that activity in the underlying temporalis musculature may thwart the interpretation of the bipolar recordings, since these potentials also show sharp gradients over both face and cortex in records taken along the entire length of this muscle.

In conclusion, we suggest that present evidence for scalp-recorded indices of cortical processes related to language production is seriously flawed due to the possible or demonstrable contamination by extracranial potentials. Since extracranial activity may precede the onset of speech sounds by at least 1 sec, no data which fail to exclude extracranial contamination through suitable recordings from electrodes over the facial region can be considered to represent valid indications of intracranial activity.

Acknowledgments

This research was carried out in part during the tenure of Exchange Scientist visits to the USA by J. SZIRTES (October, 1971 to April, 1972) and to Hungary by H. VAUGHAN (March, 1973) under the joint auspices of the Hungarian Academy of Sciences and the National Academy of Sciences. The work received support from Grants MH 06723 and HD 01799 from the USPHS. A preliminary version of this report was read at the 8th International Congress of Electroencephalography and Clinical Neurophysiology, 1973.

Prof. H.G. VAUGHAN, jr., Departments of Neurosciences and of Neurology, Albert Einstein College of Medicine of Yeshiva University, 1300 Morris Park Avenue, *Bronx, NY 10461* (USA). Tel. (212) 430 2468.

Event-Related Potentials as Signs of Sensory Processing in Left and Right Hemispheres

Language and Hemispheric Specialization in Man: Cerebral ERPs.
Prog. clin. Neurophysiol., vol. 3, Ed. J. E. DESMEDT, pp. 127–139 (Karger, Basel 1977)

Spectral Analysis of Evoked Potential Asymmetries Related to Speech Dominance

ALAN E. DAVIS and JUHN A. WADA

Division of Neurological Sciences, University of British Columbia, Vancouver, B.C.

Spectral analysis of evoked potentials (EPs) can be used to determine the amplitude characteristics of single responses and the form similarities between responses. The analysis is done separately for different frequency components of the waveforms, rather than at different peaks or other time-related components [SPEKREIJSE et al., 1977]. The amplitude characteristics of a waveform can be represented by its power or autospectrum. DAVIS [1973] has used the autospectrum of VEPs and AEPs to show two major frequency components, a 0–6-Hz band and a 6–12-Hz band. The 6–12-Hz band was of significantly greater amplitude over the occipital areas for VEPs, and over the temporal areas for AEPs. The distribution of the 0–6-Hz band was not significantly different for flash or click. The suggestion was made that the 0–6-Hz component represented a modality non-specific component, and the 6–12-Hz band a modality-specific component of the VEP and AEP.

The similarity of form between two waveforms can be measured by another spectral quantity, the coherence. DAVIS and WADA [1974] demonstrated 6–12-Hz coherence asymmetries in VEPs and AEPs which were related to the speech dominance of patients whose speech dominance had been ascertained by the carotid amytal test [WADA, 1949; WADA and RASMUSSEN, 1960]. The coherence between occipital and temporal areas was greater on the speech-dominant side for AEPs, and on the speech-non-dominant side for VEPs. These results also applied to right-handed, presumably left speech-dominant, normal subjects.

The coherence asymmetries were produced using simple flash and click stimuli with no verbal content. This indicated that hemispheric asymmetries existed at non-verbal levels, and for both auditory and visual modalities. It was proposed that the lateralized coherent processes could be related to perception, since behavioral evidence suggested that auditory and visual perception were also lateralized to the speech-dominant and non-dominant hemispheres, respectively.

The above report had several shortcomings. Coherence was calculated by a novel procedure whose statistical properties had not been studied in detail. The question of different asymmetries between the recording areas was only dealt with qualitatively. There is no obvious reason why only asymmetries between homologous, right and left hemisphere areas should be studied. For example, MISHKIN [1962] proposed a preferential, left-occipital to right-temporal pathway to explain observed visual and auditory deficits due to right and left temporal lesions, respectively.

This report extends and validates our earlier findings. Both amplitude and form asymmetries are considered using completely conventional procedures. The asymmetries are analyzed with statistical techniques which allow comparisons between all possible asymmetries of the recording areas. Using the speech dominance ascertained patients as a reference population, a simple computer test for the EP determination of speech dominance is developed. This test is then applied to 2 groups of pure right or left-handed, normal subjects to investigate the relationships of speech dominance and handedness.

Methods

The experimental format has been reported [DAVIS and WADA, 1974] and will only be briefly outlined here. The epileptic patients, 6 left speech-dominant (LSD) and 5 right speech-dominant (RSD), 8 male and 3 female, ages 13–47, were all being evaluated for surgical treatment of epileptic seizure disorders. Speech dominance was ascertained by the carotid amytal test, using femoral catheterization, and repeated on both sides at least 2 days apart.

The normal subjects were university students, selected on the basis of 13 handedness criteria [ANNETT, 1967; DAVIS and WADA, 1974]. Only 'pure' right-handed subjects who performed at least 12 out 13 tasks with their right eye, hand or leg, and similarly defined 'pure' left-handed subjects were used in the study. The final normal subject group included 12 'pure' right-handed persons, 7 male and 5 female, and 12 'pure' left-handed persons, 6 male and 6 female.

Flash stimuli were presented through a 6-inch aperture, 24-inches from the subject's nasion. Click stimuli were presented through headphones worn by the subject. Stimuli were randomly mixed and appeared at an average inter-stimulus interval of 3 sec, minimum of 2 sec. Four groups of 40 flash and 40 click stimuli were presented with a 1-min rest between groups. Monopolar recordings, referenced to linked right and left earlobes, were sampled at 100/sec and averaged into one VEP and one AEP, 500-msec duration, from each of the 4 recording sites, and each consisting of the average response to 160 stimuli.

Electrode locations were the left and right occipital (01 and 02 on the international 10-20 system) and left and right temporal (midway between T3 and C3, and midway between T4 and C4, and labelled T3 and T4, respectively).

Spectral analysis. The first 500-msec (50 data points at 100/sec sampling rate) were spectrum-analyzed using the UCLA Biomedical computer program BMD 03T – 'time series spectrum estimation' [DIXON, 1973]. Limitations of the program allowed only resolutions or bandwidths of 3, 6 or 12 Hz. In this report, the 6-Hz spectral estimate, extending over the frequency range 3–9 Hz was used for all analyses. This range was the closest overlap to the 6–12-Hz band, with peak power at about 8 Hz, studied earlier by us [DAVIS, 1973]. It was also the largest amplitude component of the entire spectra for 500-msec EPs. It could thus be used to calculate single estimates of the coherence and auto-spectra for the dominant, or largest amplitude, frequency component of the EPs. Later analyses of the coherences and autospectra required normal distributions. Accordingly, the coherences were arctanh transformed and the autospectra $\log_{10}$ transformed [JENKINS and WATTS, 1968; HANNAN, 1970]. All further discussion of coherences and autospectra will refer to their normalized forms.

Asymmetry calculations. Both coherence and autospectral asymmetries were analyzed. From the 4 recording areas, left and right occipital (01 and 02), and left and right temporal (T3 and T4), the 6 possible coherence pairs were: (01:02); (01:T3); (01:T4); (02:T3); (02:T4), and (T3:T4). From these, 3 independent coherence asymmetries could be calculated: (01:02-T3:T4); (01:T3-02:T4), and (01:T4-02:T3). Autospectral asymmetries were calculated between homologous areas of each hemisphere, and were (01–02) and (T3–T4). Since responses to both flash and click stimuli were available, the final analyses included 6 coherence asymmetries and 4 autospectral asymmetries.

Discriminant analysis. Using a technique known as discriminant analysis [SNEDECOR and COCHRAN, 1967], it was possible to determine which of the coherence or autospectral asymmetries defined above was most related to speech dominance. Equations were derived which gave quantitative measures of how much each asymmetry contributed to determination of speech dominance, and the significance of that contribution. These equations were based on the known speech dominance of the epileptic patients and could then be applied to the normal right or left-handed subjects to predict their speech dominance.

Beside providing significance levels for each asymmetry, the discriminant analysis also gave a multivariate U-statistic, similar to the F-statistic of univariate analysis of variance. This statistic was used to test the significance of the differences between the asymmetries of the left and right speech-dominant groups. The procedures for discriminant analysis are incorporated in the UCLA Biomedical computer program BMD 07M–'step-wise discriminant analysis' [DIXON, 1973]. The program was set to eliminate any asymmetries which were not related to speech dominance at a level of <0.05.

Results

Table I shows the discriminant analysis results for the 3–9-Hz coherence asymmetries. There were two significant discriminators ($p < 0.05$) which could be used to predict speech dominance. These were the left minus right hemisphere, occipital to temporal coherence (01:T3–02:T4), for flash and click. These alone could be included in an equation which correctly classified 5/6 LSD and 5/5 RSD patients on the basis of their coherence asymmetries. The probabilities of misclassification were $p < 0.05$ in 6/6 LSDs and 3/5 RSDs.

The discriminant equations were:

$$S_{LSD} = -1.79 - 5.92 \, (F) + 14.67 \, (C),$$
$$S_{RSD} = -1.14 + 1.93 \, (F) - 11.59 \, (C), \tag{1}$$

where S_{LSD} and S_{RSD} are the LSD and RSD scores, respectively, and F = flash coherence asymmetry (01:T3–02:T4), and C = click coherence asymmetry (01:T3–02:T4). Using these equations, an unknown subject is assigned to the group for which he attains the most positive score.

The posterior probabilities, P_{LSD} and P_{RSD}, that a subject has been misclassified are:

$$P_{LSD} = \frac{1}{1 + \exp(S_{LSD} - S_{RSD})}$$
$$P_{RSD} = \frac{1}{1 + \exp(S_{RSD} - S_{LSD})} = 1 - P_{LSD}. \tag{2}$$

For the significant coherence asymmetry (01:T3–02:T4), the U-statistic showed that the LSD and RSD means were significantly different at $p < 0.005$.

Table II applies the above discriminant equations to the coherence asymmetries of the normal subjects. A significance level of $p < 0.05$ for the posterior probability was used to definitely assign speech dominance. Probabilities of > 0.05 were classified as bilateral. Table II shows that for the 12 right-handed subjects, 8 were LSD, 3 RSD, and 1 bilateral. For the 12 left-handed subjects, 5 were LSD, 1 RSD, and 6 bilateral. In the bilateral subjects, the single right-handed subject was more RSD than LSD, and 4 of the 6 left-handed subjects were more LSD than RSD.

The mean power spectra of all subjects showed that the 6-Hz point, with a frequency range of 3–9-Hz, contained from 65 to 75% of the total power of the response. Table III shows the discriminant analysis of this auto-

Table I. Discriminant analysis of coherence asymmetries

Coherence asymmetry	LSD mean (6 patients)	RSD mean (5 patients)	Discrimination probability
(01:02–T3:T4)F	0.02	0.08	0.286
(01:02–T3:T4)C	−0.18	−0.06	0.243
(01:T3–02:T4)F	−0.34	−0.19	0.037 (p<0.05)
(01:T3–02:T4)C	0.11	−0.24	0.001 (p<0.005)
(01:T4–02:T3)F	−0.36	0.09	0.240
(01:T4–02:T3)C	−0.05	−0.17	0.941

01, 02 = left, right occipital; T3, T4 = left, right temporal. F, C = flash, click. Discrimination probability is used to measure the significance of each asymmetry's relationship to speech dominance.

Table II. Distribution of speech dominance, using discriminant equations of coherence asymmetries

Subject group	Number of subjects	Coherence asymmetry classification		
		LSD	RSD	BIL
LSD amytal	6	6	0	0
RSD amytal	5	0	3	2 (2 RSD)
Right handed	12	8	3	1 (1 RSD)
Left handed	12	5	1	6 (2 RSD)

LSD, RSD = Left, right speech dominant; BIL = bilateral. Subjects were classified as LSD or RSD if their classification probabilities were less than 0.05, and BIL if greater.

Table III. Discriminant analysis of autospectral asymmetries

Autospectral asymmetry	LSD mean (6 patients)	RSD mean (5 patients)	Discrimination probability
(01–02)F	−0.048	−0.074	0.763
(01–02)C	−0.043	−0.049	0.879
(T3–T4)F	−0.106	−0.081	0.848
(T3–T4)C	−0.115	−0.017	0.474

01, 02 = Left, right occipital; T3, T4 = left, right temporal; F, C = flash, click. Discrimination probabilities are used to measure the significance of each asymmetry's relationship to speech dominance.

Table IV. Analysis of variance of left minus right hemisphere, occipital to temporal coherence (01:T3–02:T4) for 12 right-handed and 12 left-handed subjects

Source	F value	F probability
Mean	2.734	0.109
Handedness	0.489	0.498
Stimulus modality	5.926	0.022 ($p < 0.05$)
Interaction	1.514	0.230

spectral or amplitude-related component of the EPs. There were no significant predictors ($p < 0.05$) of speech dominance, and the U-statistic showed no significant differences ($p < 0.05$) between the LSD and RSD group means. Discriminant analysis of the autospectra, themselves, rather than the differences between homologous areas, also gave insignificant results.

Significant amplitude asymmetries could be demonstrated by grouping all the normal subjects together, regardless of their handedness. The right hemisphere ($02 + T4$) autospectra were greater than the left ($01 + T3$) for 19/24 VEPs, which was significant at $p < 0.001$ with the Mann-Whitney, non-parametric U-test. For the AEPs, 16/24 were greater on the left side (almost significant at $p < 0.05$).

The (01:T3–02:T4) coherence asymmetry was the only significant discriminator of speech dominance. A univariate analysis of variance of this asymmetry as the dependent variable was used to study the effects of stimulus modality and handedness. The flash and click stimuli were repeated measures performed upon the same subject. Subjects (S) were nested in handedness (F) and crossed with stimulus modality (M). Table IV shows the results of the analysis of variance for the 24 normal subjects, using the UCLA Biomedical computer program BMD 08V – 'analysis of variance' [DIXON, 1973]. The flash and click means were significantly different ($p < 0.05$), but the left and right-handed means were not. In addition, there was no significant interaction between handedness and modality.

Discussion

Coherence and autospectra have fairly simple interpretations in terms of the amplitude and form of EPs. The autospectrum measures the amount of energy or amplitude of each frequency component of the EP, while coherence measures the amount of similarity of form between two EPs. Coherence is similar to correlation except that it measures similarity between two waveforms after they have been aligned so their correlation is maximized in each frequency range. Thus, 2 waveforms of very similar profile, but of different latencies, will be uncorrelated but highly coherent.

The results of the discriminant analysis of the coherence or form-related asymmetries showed a significant relationship to speech dominance, since the mean asymmetries were significantly different for the LSD and RSD groups. In contrast, the autospectra or amplitude-related asymmetries did not appear to be consistently related to speech dominance.

Amplitude asymmetries of event-related potentials have been reported for natural speech stimuli [MORRELL and SALAMY, 1971], during linguistic versus non-linguistic tasks [WOOD, this volume], with meaningful versus non-meaningful stimuli [MATSUMIYA et al., 1972], for tachistoscopic presentations of words and similarly patterned stimuli [BUCHSBAUM and FEDIO, 1970] and for motor potentials preceding articulation [MCADAM and WHITTAKER, 1971]. In a lower frequency range, MORRELL and HUNTINGTON [1971] and Low et al. [1973] have reported slow, CNV-type, potential amplitude asymmetries preparatory to speech.

Our negative results for amplitude asymmetries related to speech dominance do not necessarily contradict the above findings. Only Low et al. used subjects whose speech dominance was known. The others suggested that the observed amplitude asymmetries of right-handed subjects might be generally related to speech dominance. Amplitude asymmetries did exist for the normal subjects. But, as will be discussed below, the coherence asymmetries of both left and right-handed subjects showed a primarily LSD pattern. The normal subject results thus neither prove nor disprove the relationship of amplitude asymmetries to speech dominance, since most are probably LSD.

The discriminant analysis of the speech dominance ascertained patients, however, shows clearly that coherence asymmetries are significantly related to speech dominance, but autospectral asymmetries are not. We therefore conclude that for simple stimuli, form asymmetries are more closely related to speech dominance than are amplitude asymmetries.

The discriminant equations 1 and 2 for the coherence asymmetries give a quantitative comparison of the contribution of the flash and click asymmetries to the determination of speech dominance. First, the LSD and RSD coefficients have opposite signs for both flash and click. This indicates that flash and click asymmetries contribute oppositely the determination of speech dominance, and that LSD and RSD scores are affected oppositely by these asymmetries. The coefficient signs show that a large AEP coherence within one hemisphere increases the score, and consequently the probability, that speech dominance is localized within that hemisphere. Conversely, a large VEP coherence in one hemisphere increases the probability that speech dominance is localized within the opposite hemisphere.

A second comparison may be made between the magnitudes of the coefficients. The approximately equal magnitudes of the click coefficients (14.67 and 11.59) for the LSD and RSD groups show that click asymmetry magnitudes are about equally important in determining LSD or RSD membership. In contrast, the smaller flash coefficients (5.92 and 1.93) for the LSD and RSD groups show that flash asymmetries are less important than click asymmetries in determining speech dominance. In addition, flash asymmetries contribute more to LSD than to RSD determination.

Davis and Wada [1974] have discussed the possible significance of these coherence asymmetries. For AEPs, they found that 4/5 LSDs, 4/5 RSDs, and 10/12 normal, right-handed subjects showed greater, occipital-temporal coherences on the speech dominant side. For VEPs, 4/5 LSDs, 3/5 RSDs, and 9/12 normals showed greater, occipital-temporal coherences on the speech-non-dominant side. It was assumed that the right-handed subjects were primarily left speech dominant. These results suggested that lateralization of hemispheric function could occur with non-verbal stimuli for both visual and auditory modalities, and that these functions were clearly separated to opposite hemispheres. In terms of occipital and temporal form similarities, the AEP forms were more similar within the speech-dominant hemisphere, while VEP forms were more similar within the speech-non-dominant hemisphere.

It was hypothesized that the coherent EPs were in some way related to associative processes of auditory and visual perception, since these functions were also lateralized to the speech-dominant and non-dominant hemispheres, respectively. Evidence for this lateralization included deficits observed with temporal lobe lesions [MOUNTCASTLE, 1962; GESCHWIND, 1970], pharmacological paralysis [WADA, 1949; WADA and RASMUSSEN, 1960], split brain humans [SPERRY, 1974; SPERRY et al., 1969, GAZZANIGA, 1970], and dichotic listening [KIMURA ,1961]. The possibility that coherent or correlated processes are related to perception is also suggested by the work of several Russian investigators. They reported that the correlations of EEG activity between many scalp areas increased during problem-solving [ASLANOV, 1970], during conditioned reflex formation [DUMENKO, 1970], and were

abnormally high in persons with associative disorders related to schizo-
phrenia and epilepsy [GAVRILOVA, 1970].

The application of the discriminant equations relating coherence asymmetries to the speech dominance of normal subjects is of great interest, since
no unequivocal method exists for this determination. Definitive tests, such
as the carotid amytal test or electrostimulation of the exposed cortex
[PENFIELD and ROBERTS, 1959], are only justified for diagnostic or therapeutic
purposes.

Compared to amytal test, the discriminant equations are approximately
90% accurate. The amytal test, itself, has been surgically verified as being
greater than 95% accurate by BRANCH et al. [1964]. Using a significance
level of $p < 0.05$ to definitely assign speech dominance, the discriminant
equations classified the 12 right-handed subjects into 8 LSD, 3 RSD, and
1 bilateral, and the 12 left-handed subjects into 5 LSD, 1 RSD, and 6 bilateral. If bilaterality is ignored, 8/12 right-handed subjects and 9/12 left-
handed subjects were LSD.

The relationships between handedness, stimulus modality and coherence
asymmetries are clearly shown in the analysis of variance results. For both
the left and right-handed subjects, the VEP coherences were significantly
greater in the right hemisphere and AEP coherences were significantly
greater in the left hemisphere. Handedness had no significant effect on
coherence asymmetries and there was no interaction between handedness
and stimulus modality. The significant modality effect, and insignificant
handedness effect and interaction show that both left- and right-handed subjects respond in the same way to the stimulus modality; i.e. larger VEP
coherences occur in the right hemisphere and larger AEP coherences occur
in the left hemisphere. These were the same coherence asymmetries used to
predict 10/11 patients correctly, and for which the LSD and RSD group
means were significantly different.

From this analysis, we conclude that handedness is not significantly
related to speech dominance. In other words, the majority of the population
is left speech-dominant, regardless of their handedness. There does appear
to be some trend in the left-handed subjects to show a greater degree of
bilateral representation than do right-handed subjects. But, if bilaterality is
ignored, approximately 70–80% of both groups are LSD.

The relationship of handedness to speech dominance has been investigated by many authors, with many differing conclusions. It has been stated
that right-handedness is a good indication of left speech dominance, but that
the speech dominance of left-handed subjects is either unpredictable or

slightly favored to the right hemisphere [BENTON, 1962; HÉCAEN and AJURIAGUERRA, 1964]. It is important to note, however, that this relationship has been based on very little direct evidence. Almost all studies involved patients in whom the speech process has been interrupted by pathological, surgical or pharmacological means. It has also been difficult to determine the extent to which early injuries or unknown conditions have affected maturation and lateralization of the speech processes. Generally, too, the question of bilateral representation has been ignored.

PENFIELD and ROBERTS [1959], using electrostimulation of the exposed cortex, concluded, as we have, that handedness and speech dominance are not related. They estimated 98% LSD and 2% RSD in the right-handed population, and 90% LSD and 10% RSD in the left-handed population. Bilaterality was not considered. BRANCH et al. [1964], using the carotid amytal test, estimated 90% LSD and 10% RSD in the right-handed population, and 40% RSD, 50% LSD and 10% bilateral in the left-handed population. They admitted, however, that the question of bilaterality only became apparent some distance through their series of patients. Low et al. [1973] have reported CNV asymmetries in a group of 8 LSD and 3 RSD carotid amytal tested patients and applied these results to speech dominance determination in normal subjects. Using maximum CNV amplitude preparatory to speech as an index of laterality, they correctly predicted the speech dominance of 8/8 LSD patients and 2/3 RSD patients.

In a normal population of 11 'pure' [ANNETT, 1967] right-handed and 11 'pure' left-handed subjects, Low et al. found a distribution of CNV asymmetries corresponding to 8 LSD, 1 RSD, and 2 bilateral in the right-handed subjects. In the left-handed subjects, the distribution was 3 LSD, 6 RSD, and 2 bilateral [see also ANNETT, 1972].

The right-handed distribution of Low et al. is similar to ours derived from the discriminant equations, but the left-handed distribution is not. The major difference, however, is in the number of RSD subjects. Since only 3 RSD amytal patients were tested with their procedure, and one of these was incorrectly classified, their RSD distribution must be evaluated cautiously. Their data also shows more bilaterality than earlier studies, which agrees with our results.

Hemispheric asymmetries may exist at a very early age, and if so, discriminant analysis of EP asymmetries may be a way of studying them. WADA et al. [1975] have found that significant morphological asymmetries exist in the cortical speech zones of both infant and adult brains. Their findings also suggest that a higher percentage of persons may have right-sided cerebral speech dominance or bilateral representation than has been previously assumed. Further, morphological findings of measurable asymmetry in cortical speech areas were seen as early as the 29th gestational week.

The results of WADA *et al.* suggest that the human brain may possess a predetermined, morphological and functional capacity for the development of lateralized speech and language function. To clarify this problem from an EP viewpoint, an ongoing, longitudinal study of neonatal hemispheric asymmetry is presently underway in our lab. Some behavioral data by EIMAS *et al.* [1971] and AEP data by MOLFESE [this volume] supports the possibility of early development of these asymmetries.

This possibility is strengthened by our studies of more than 50 human infants of less than 3 months of age [DAVIS and WADA, 1977a; WADA and DAVIS, 1977]. In the majority of these infants, coherences were larger in the left hemisphere for click AEPs and in the right hemisphere for flash VEPs – the same pattern as seen in adults. The presence of these hemispheric asymmetries in pre-verbal infants and in response to non-verbal stimuli strongly suggests that language is not the basis of the most fundamental form of hemispheric specialization.

There are some problems with the application of spectral analysis, and consequently the discriminant equations, to the investigation of single subjects. Most of these problems arise because the short time duration of EPs makes the spectral estimates highly variable. Intuitively, this means that in one EP, only a few oscillations are available for analysis, and so only a small amount of reliable information can be extracted. This is in contrast to the more conventional application of spectral analysis to EEG in which long data records of many seconds are available. Short EP records thus generate individual spectra which are highly variable both within and between subjects. Consequently, coherences and autospectra calculated from these spectra are also highly variable and subject to some reservation.

The variability of coherences and autospectra, however, is not a problem in the analysis of group tendencies. In this case, each coherence or autospectral estimate is taken as a single observation, and the techniques of analysis of variance are applied. The conclusions based on group analysis are thus much more reliable than those based on single subjects. Reliability thus extends to all analysis except the distribution of speech dominance in the normal subjects. Since those distributions were based on individual subject spectra, they are not as accurate as the group analyses. The trends within those distributions, however, should be approximately correct.

Bearing in mind the variability of individual responses, a fairly simple, computational procedure for the determination of speech dominance can be defined. The BMD programs are widely available, but any spectral analysis program could be used to calculate the coherence asymmetries. Speech dominance classifications and probabilities can then be calculated using equations 1 and 2.

Elsewhere we have reported more detailed statistical analyses of this group of patients [DAVIS and WADA, 1977d] and coherence asymmetries in a larger group of 22 amytal patients [DAVIS and WADA, 1977b]. The relationships between handedness, sex and speech dominance and their effects on coherence and amplitude asymmetries have also been investigated [DAVIS and WADA, 1977c]. These reports confirm the present conclusion that lateralization of speech dominance primarily affects coherence rather than amplitude asymmetries. Furthermore they show that both handedness and sex affect EP amplitude asymmetries.

On the basis of these results, DAVIS and WADA proposed that the left and right hemispheres may be specialized for the processing of temporally and spatially structured information, respectively. To test this hypothesis, DAVIS and CRAPPER [1977] presented either visual or auditory stimuli as a sequence or as a single group. With either lines presented on an oscilloscope or tones presented through earphones, a sequence of stimuli in one modality increased response coherences in the left hemisphere and decreased them in the right. DAVIS and CRAPPER concluded that a primary difference between the left and right hemispheres was the ability to process temporal and spatial structure. A further analysis of spectral phase relationships provided a functional basis for this specialization.

In conclusion, the discriminant analysis of the spectral quantities has shown several important points. First, EP coherence or form similarities are significantly related to speech dominance, but amplitudes are not. Second, of the coherence asymmetries, only the left minus right hemisphere difference is significantly related to speech dominance. Third, the coherence asymmetries of normal left and right-handed subjects do not differ significantly. Both conform to a left speech dominant pattern; i.e. AEP coherences are significantly larger in the left hemisphere, and VEP coherences are significantly larger in the right. There did seem to be some trend to more bilaterality in left-handed subjects. Fourth, these results have been established for the largest amplitude, frequency range (3–9-Hz) of the 500-msec EP. This range approximately overlaps the modality dependent frequency components investigated by DAVIS [1973] and DAVIS and WADA [1974].

Summary

Spectral analysis of visual and auditory evoked potentials (VEPs and AEPs) was done for the 3–9-Hz, maximum amplitude, frequency component of 500-msec EPs. Responses to flash and click stimuli were recorded from the left and right, occipital and temporal areas. The subjects were 6 left and 5 right speech dominant (LSD and RSD) epileptics whose speech dominance had been ascertained by the carotid amytal test, and 12 'pure' right-handed and 12 'pure' left-handed, normal subjects. Discriminant analysis of the spectral analysis results showed that form asymmetries, as represented by the coherences, were significantly related to speech dominance, but auto-spectral or amplitude asymmetries were not. Large AEP coherences or similarity of form between the occipital and temporal responses of one hemisphere increased the probability that speech dominance was localized within that hemisphere. Conversely, similar large VEP coherences within one hemisphere increased the probability that speech dominance was localized within the opposite hemisphere. Discriminant equations using this coherence asymmetry correctly predicted the speech dominance of 10/11 patients. In the normal subjects, the equations classified the 12 right-handed subjects into 8 LSD, 3 RSD, and 1 bilateral, and the 12 left-handed subjects into 5 LSD, 1 RSD, and 6 bilateral. Analysis of variance showed that both left- and right-handed subjects had significant, LSD coherence asymmetries; i.e. occipital-temporal coherences were significantly larger in the left hemisphere for AEPs, and in the right hemisphere for VEPs. This suggested that handedness and speech dominance were not related, although left-handed subjects showed some trend to more bilaterality. The discriminant analysis of coherence asymmetries is a simple basis for the computational, EP determination of speech dominance in normal subjects and patients.

Acknowledgements

We wish to thank ANNE HAMM and ED JUNG for expert technical assistance, and the Department of Physiology, UBC for computing funds. This work was supported by US NIH and Canadian MRC grant to JAW. AED is a MacMillan Family Fellow.

Prof. JUHN A. WADA, Division of Neurological Sciences, Health Sciences Center Hospital, 2075 Wesbrook Place, *Vancouver, B.C. V6T 1W5* (Canada). Tel. (604) 228 2012.

Language and Hemispheric Specialization in Man: Cerebral ERPs.
Prog. clin. Neurophysiol., vol. 3, Ed. J.E. Desmedt, pp. 140–150 (Karger, Basel 1977)

Indices of Lateralized Cognitive Processes:
Relation of Evoked Potential Asymmetry to
EEG Alpha Asymmetry

David Galin and Ron R. Ellis

Langley Porter Neuropsychiatric Institute, University of California, San Francisco, Calif.

Hemispheric Specialization for Different Cognitive Modes

Our understanding of lateral cerebral specialization is based very largely on studies of behavioral deficits produced by unilateral lesions Milner, 1971; Luria, 1966; Zangwill, 1960] and cerebral disconnections [Sperry, 1973; Geschwind, 1965] and, only recently, on studies of normal subjects [Galin and Ornstein, 1972; Doyle *et al.*, 1974]. It is generally agreed that in typical right handers language processes and arithmetic depend primarily on the left hemisphere, and that the right hemisphere is particularly specialized for spatial relations and some musical functions. The situation with left handers is more complex, and has been summarized by Hécaen [1964], Silverman *et al.* [1966], James *et al.* [1967], Levy [1969] and Miller [1971]. For example, right temporal lobectomy produces a severe impairment on visual and tactile mazes; left temporal lesions of equal extent produce little deficit on these tasks, but specifically impair verbal memory. In the 'split-brain' patients, the left hemisphere is capable of speech, writing, and calculation, but is severely limited in problems involving spatial relationships and novel figures. The right hemisphere can easily carry out tasks involving complex spatial and musical patterns, but can perform simple addition only up to 10, and has the use of relatively few words – very little expressive language and comprehension of syntax at about the level of a 2-year-old [Gazzaniga and Hillyard, 1971; Levy *et al.* 1971]; see Smith [1966] for a report of modest recovery or relearning of speech following left hemispherectomy.

It is important to emphasize that what most characterizes the hemispheres is not that they are specialized to work with different types of material, the left with words and the right with spatial forms. Rather each hemisphere is specialized for a different cognitive style: the left for an analytical, logical mode for which words are an excellent tool, and the right for a holistic, gestalt mode which happens to be particularly suitable for spatial relations, as well as music.

Some caution should be exercised in making the inference of lateral specialization of cognitive function in normal people from lesion studies alone. One might consider whether the 'split' functions are due in some part to the radical surgery, or to the other disturbances in these patients. The study of neurological disorders or surgical preparations cast light on normal functioning, but the most important and most practical question is whether the normal brain, engaged in everyday activities, is organized around lateralization of cognitive function.

Therefore, it is of great importance that this lateralization of cognitive function has been demonstrated in normal subjects with EEG techniques [GALIN and ORNSTEIN, 1972; McKEE et al., 1973; DOYLE et al., 1974; SCHWARTZ et al., 1973], evoked potentials [BUCHSBAUM and FEDIO, 1969, 1970; MATSUMIYA et al., 1972; WOOD et al., 1971; MORRELL and SALAMY, 1971; VELLA et al., 1972], and with studies of left/right visual field differences in perception and reaction time [WHITE, 1969; FILBEY and GAZZANIGA, 1969; McKEEVER and HULING, 1970; RIZZOLATTI et al., 1971; BERLUCCHI et al., 1971], and with asymmetries in 'reflective' eye movements [KOCEL et al., 1972; GALIN and ORNSTEIN, 1974; KINSBOURNE, 1972].

For example, we have examined the EEGs of normal subjects performing verbal and spatial tasks to determine whether there were differences in activity between the two hemispheres [GALIN and ORNSTEIN, 1972; DOYLE et al., 1974]. We found relatively higher alpha amplitude (which we interpret as a measure of idling) over the right hemisphere during the verbal tasks, and relatively more alpha over the left hemisphere during the spatial tasks. In other words, the hemisphere expected to be less engaged in the task has more of the idling rhythm. Other investigators have confirmed our observations on changes in alpha asymmetry in studies which contrasted verbal and musical tasks [McKEE et al., 1973; SCHWARTZ et al., 1973]. Other laboratories using the evoked potential (EP) method have reported that EPs to speech sounds were larger in the left hemisphere than in the right [WOOD et al., 1971], and that EPs to complex visual forms were larger on the right than on the left [VELLA et al., 1972]. Another large group of studies with normal subjects have made use of the different hemispheric projections of the left and right visual fields and have found, for example, that the left and right hemispheres have opposite superiorities in discriminative reaction times to stimuli such as letters and faces [RIZZOLATTI *et al.*, 1971; BERLUCCHI *et al.*, 1971].

In this experiment, we sought to compare the asymmetry of averaged visual EPs and the EEG alpha power as indicators of lateralized cognitive functions. Although there have been several reports that EP asymmetry is related to lateral specialization [BUCHSBAUM and FEDIO, 1969, 1970; WOOD *et al.*, 1971; MORRELL and SALAMY, 1971; VELLA *et al.*, 1972; MATSUMIYA *et al.*, 1972], to our knowledge, there has been no direct comparison of asymmetry of alpha and EP in the same subjects, performing the same tasks.

In this experiment, we manipulated the EEG alpha asymmetry and observed the effect on superimposed EP. We did this by engaging the subjects in spatial and verbal tasks, and simultaneously presenting flashes throughout the performance of the task.

Method

Six right-handed adults aged 20–40 served as subjects (4 male, 2 female). The subject was seated in a straight-backed chair at a small table in a shielded sound-attenuating room. Reliability of electrode placement was assured with a standardized electrode cap with fittings to hold electrodes at the 10–20 system positions. Recordings were made from P3, P4, T3, and T4, all referenced to Cz.

The considerations in choice of reference site are discussed by GIBBS and GIBBS [1952] and COOPER *et al.* [1974]. A 'common reference' recording (monopolar) is more suitable for detecting asymmetry between homologous scalp locations than bipolar recording; any difference between the two homologous channels must be due to activity at the 'exploratory' lead. We have chosen Cz as reference electrode rather than the more usual ear reference because the ear carries a large component of the temporal lobe signals, which we expect to be particularly sensitive to the different cognitive conditions we are studying. The linked-ears reference can lead to misinterpretation of localization in cases where the temporal lobe signal is asymmetrical [GIBB and GIBBS, 1952, vol. 2, pp. 165–166].

Vertical and horizontal EOGs were recorded to facilitate editing the record for artifact. Electrodes were attached at the outer canthus of each eye, and above and below the left eye. Recordings were made with a Grass model 7 polygraph (1/2 amplitude cutoff = 1–35 Hz) and a 4 channel Hewlett Packard FM tape recorder augmented to 7 channels with a Vetter multiplexer. Recordings were made while the subject performed a series of 3-min tasks: a practice trial was given each type of task. To prevent habituation from confounding task differences in the EP, the tasks were given twice in an ABBA order with task order counterbalanced across subjects.

a) Modified Kohs block design: In this task, the subject was given a two-dimensional geometrical pattern to memorize for 1 min and was then asked to construct the pattern with a set of 16 multicolored blocks.

b) Writing from memory: In this task, the subject was given 1 min to read a passage containing minimal imagery taken from the newspaper and then was asked to write from memory as many of the facts and ideas as he could in 3 min.

Two 12-inch fluorescent lamps (Iconix model No. 6109-9 hot cathode) were placed 32 inches apart, 28 inches in front of the subject, one on each side of the work table. The lamps were powered by an Iconix lampdriver (model No. 6196-2), producing one 10-msec flash every 3 sec continuously throughout the session. Lampdriver 'intensity' control was set at 7.5; background light intensity was 210 lux, flash intensity was 420 lux.

The recordings from each lead were processed off-line by a PDP-7 computer for Fourier spectrum analysis and computation of average alpha power ratios between homologous leads over the entire 3-min task [for details, see DOYE et al., 1974], and for computation of averaged EP for 500 msec following each flash. All trials on which blinks or eye movements greater than 6° occurred were omitted from the averages. For all subjects at least 30 flash EPs were averaged for each of the tasks (blocks I and II; writing I and II).

Several characteristics of the EPs were examined. The simplest was a power measure which reflects the overall amplitude of the EPs without considering details of the waveform. The standard deviation squared was taken as an approximation to the whole band power of the averaged waveform [CALLAWAY and HALLIDAY, 1973]. This was computed for the interval 50–450 msec, which encompassed the main features of the EPs, and for the segment 100–350 msec, which contained the components which appeared to be most variable across tasks within subject, and therefore might be most task-dependent.

In addition, we attempted to define individual components in the waveform and describe their peak amplitude and latency. Peak identification in EPs is subject to many methodological problems. Although there is enough consistency in EP waveform across subjects to encourage some investigators to identify components by polarity and latency, the variability even within subject and within condition is great. In our small sample, we found components that appeared clearly at the same latency in each condition in some subjects, but were absent in others, and even within subject, present in one cognitive condition and absent in the other. With this caveat, however, we include these data in our comparison, since peak measures are among the most commonly used to describe EPs.

Peaks of individual components were selected as follows: for each subject, the plotted EPs for each homologous pair of leads (e.g. P3, P4) over all task conditions were visually examined by both authors, and those peaks which appeared in all or nearly all of the plots were noted. The range of latencies of the clearly defined peaks was used as a guide in determining the appropriate point to be taken in the more ambiguous plots. In this way, in almost every case corresponding peaks could be located in each condition. In some cases, however, no trace of a component could be found, i.e. no maximum or minimum occurred within 50 msec of the appropriate latency; the value for the amplitude of that component on that condition was taken as the value of the plot at the mean latency for that peak in other conditions. In this manner, the peaks to be compared across conditions for each pair of leads in each subject were identified.

The amplitude of the peak-trough difference of successive components was computed, e.g. from the trough of the first negative wave to the peak of the first positive wave, and from the first positive to the second negative, etc. For each task, for each pair of homologous leads, a right/left ratio score comparable to the EEG alpha power ratio was computed for the measures of EP power and for the peak-trough amplitude of individual components. The ratios for blocks I and II, and for writing I and II, were combined by computing the geometric mean.

Results

Illustrative EPs from one subject (No. 1a) are shown in figure 1. Fourier power spectra of this subject's EEG during each task episode are shown in figure 2. Right/left ratios for each subject are shown in table I.

1. EEG alpha power ratios. For all leads in all subjects, the alpha power ratio right/left was higher in the write-from-memory task than in the blocks design task. This measure appeared stable; variability in ratio values from task I to task II was usually less than 0.15. This stability is well illustrated in

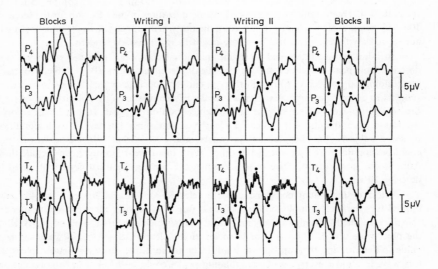

Fig. 1. Evoked potentials from subject 1a (table I). Right/left ratio of overall power and amplitude of individual components is higher in verbal tasks than in spatial tasks. P4 = Right parietal; P3 = left parietal; T4 = right anterior temporal; T3 = left anterior temporal. All leads referred to vertex (Cz). Vertical lines spaced 100 msec apart. Dots indicate peaks selected for the amplitude measures. Polarity = positive up.

figure 2; very similar spectra are seen when the tasks are repeated. Within-subject reliability was checked by comparing writing I to blocks I, writing II to blocks II; in 27 of the 28 available comparisons (all subjects, leads), the writing ratio was higher than the blocks ratio. The alpha power averages were also computed with the epochs containing the flash and EP edited out; this did not change the ratios by more than 0.07.

2. *EP power ratio.* With a few exceptions, the results paralleled the EEG alpha power measure; EP power right/left (100–350 msec) was greater during writing than blocks in all subjects in all leads except subject No. 2. Variability from task I to II was greater than with EEG alpha power. Results using the longer interval 50–450 msec were not as consistent.

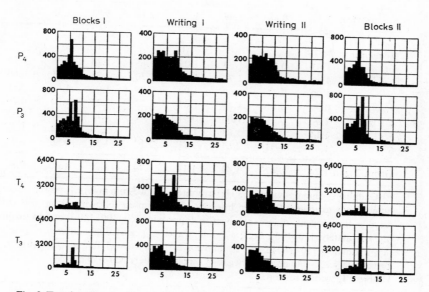

Fig. 2. Fourier power spectra for blocks and writing from memory tasks for subject 1a. These spectra were computed over the entire 3-min task periods during which the flash-evoked responses in figure 1 were elicited. For each lead EEG power is plotted versus frequency in 1-Hz intervals from 1 to 29 Hz; the last point on each plot is an average for frequencies 30–64 Hz. The ordinate is scaled in arbitrary units in which a 10-Hz sine wave of 80 V p-p corresponds to 80,000 units. The ratio of alpha-band power from homologous leads T4/T3 and P4/P3 is smaller on the blocks task than on the writing task. Note change of ordinate scale between tasks. This is necessary to accommodate high left alpha peaks during the block design. This subject also shows high but generally symmetrical theta peaks which we often observe during this task. For details on the behavior of the other frequency bands, see DOYLE *et al.* [1974].

Table I. Right/left ratios. Entries for each subject represent the geometric means of ratios for writing I and II and for blocks I and II

Subjects	EEG alpha		EP power		EP early wave		EP late wave	
	writing	blocks	writing	blocks	writing	blocks	writing	blocks
Parietal = P4/P3								
1a	1.55	0.60	1.93	1.41	3.07	2.10	1.01	0.98
1b[1]	1.53	0.54	3.29	1.74	2.80	2.05	1.20	1.01
2	1.13	0.82	0.85	0.88	1.44	1.15	0.87	0.91
3	1.23	1.17	2.02	1.63	7.04	5.58	1.15	1.09
4	1.06	0.91	1.42	0.83	0.99	1.18	1.24	0.98
5	1.23	0.87	2.60	0.82	1.21	0.84	3.07	1.90
6	1.94	1.60	3.06	1.78	2.21	1.70	1.84	1.26
Geom. Mean	1.35	0.88	1.99	1.23	2.18	1.73	1.36	1.13
Temporal = T4/T3								
1a	1.63	0.37	1.39	1.03	1.68	1.26	0.88	0.75
1b[1]	1.69	0.42	1.71	1.32	2.30	1.35	0.91	0.80
2	1.00	0.81	0.69	1.00	0.88	1.01	1.00	0.81
3	0.99	0.74	1.43	1.09	1.22	1.22	1.00	0.96
4	1.00	0.81	1.06	0.94	1.11	1.11	0.99	0.85
5	1.14	0.46	1.88	1.05	1.29	1.23	1.44	1.31
6	1.47	1.07	2.65	1.71	2.12	1.52	0.86	0.97
Geom. Mean	1.24	0.62	1.43	1.14	1.44	1.23	1.00	0.91

EEG alpha = alpha power (8–13 Hz) averaged over a 3-min task; EP power = computed from averaged EP, over interval 100–350 msec following flash; EP early wave = trough-peak difference negative (approximately 100–140 msec) to positive (approximately 150–180 msec); EP late wave = peak-trough difference, positive (approximately 150–180 msec); EP late wave = peak-trough difference, positive (approximately 250–280 msec) to negative (approximately 320–350 msec).
[1] Subject 1, second session.

3. EP peak-trough amplitude. In general, the peak measures paralleled the EEG alpha power measure, but were much more variable, within and across subjects. In the EPs from the parietal leads, all subjects had an early negative (100–140 msec)-positive (150–180 msec) configuration and a later positive (250–280 msec)-negative (320–350 msec) configuration. For both the early and late waves, 5 of 6 subjects had higher ratios during writing than during blocks.

Another negative trough (180–210 msec) between these two was clearly defined in 4 of the 6 subjects (fig. 1). Right/left ratios of peak-trough differences involving this component were not consistently task-dependent.

The EPs from the anterior temporal leads T3 and T4 were even more variable than from the parietals. As in the parietals, an early and a late component could be defined. The ratio for the peak-trough differences of the late component (230–280 msec positive; 300–350 msec negative) were higher during writing than during blocks in 5 of 6 subjects; for the early component (100–140 msec negative; 150–220 msec positive) 3 were higher and 3 were lower or equal.

4. Latency. Latencies of the peaks showed no systematic changes between writing and blocks in either the temporal or parietal leads. In all tasks, however, the early negative component (100–140 msec) had longer latencies on P3 and T3 than on P4 and T4 (5 of 6 subjects). The late positive component (250–280 msec) had longer latency at P3 than P4 in 6 of 6 subjects, but was not consistent in the temporal leads.

5. Correlation between task dependence of EP and EEG alpha power. The change in ratio from writing to blocks was computed for EEG alpha and the EP measures. Rank order correlations between the change score for alpha power and for each of the EP measures were computed. The correlations between these indices in the temporal leads were much higher than in the parietal leads:

Alpha vs EP power:	Temporal r = 0.50	Parietal r = 0.29
Alpha vs EP early wave:	Temporal r = 0.79	Parietal r = 0.18
Alpha vs EP late wave:	Temporal r = −0.43	Parietal r = −0.14

Discussion

The results indicate that overall power and peak amplitude characteristics of EP asymmetry reflect the lateralization of cognitive processes, but not as consistently as the concomitant asymmetry in EEG alpha power. The EP power measure appeared less variable and more task-dependent than the individual peak amplitude measures. The parietal and anterior temporal leads behaved similarly under these conditions, with the exception that the task-dependence of EEG alpha asymmetry was more clearly correlated with the task dependence of the EPs in the temporal than in the parietal leads.

Whether the right/left ratio is above or below unity may not be significant by itself. As LEISSNER *et al.* [1970] have pointed out, normal asymmetries in skull thickness of up to 30% are common, and highly correlated with EEG asymmetry. Therefore, we are concerned here with the *change* in alpha ratio from one cognitive task to another.

In this study, the stimulus for the EP was irrelevant to the task, i.e. the flash was used as a neutral probe of hemispheric excitability. Several previous studies have shown that under these conditions the amplitude of the EPs increases as a function of the background EEG alpha [LANSING and BARLOW, 1972; SPILKER *et al.*, 1969]. In this respect, our experiment is quite different from others which have examined the EP asymmetry in relation to lateral specialization, using stimuli which were central to the task, e.g. discrimination of words or geometric forms [BUCHSBAUM and FEDIO, 1969; MORRELL and SALAMY, 1971; VELLA *et al.*, 1972]. When the stimulus is task-relevant, the EP amplitude increases as a function of increasing attention or involvement [DONCHIN and LINDSLEY, 1969; DESMEDT, 1977c], conditions usually associated with decreasing background EEG alpha. This difference must be kept in mind when comparing these 2 types of studies. Another difference is that in our present experiment, we are not concerned with the absolute amount of alpha activity at any one location, but rather with the relative distribution of alpha over the left and right hemispheres.

Several reports have linked EP asymmetry with intelligence [CALLAWAY, 1973; CALLAWAY and STONE, 1969; RHODES *et al.*, 1969]. In the testing procedure of most of these reports, the subjects were either 'at rest' or minimally constrained by the cognitive demands of the test situation. Under these circumstances, the subjects may assume their preferred cognitive mode; e.g. some daydreaming in vivid imagery, others rehearsing conversations. We suspect that asymmetries in EP amplitude might be in part dependent on asymmetries in the alpha amplitude of the background EEG, which we have shown can be a function of which cognitive mode the subject is using. Their EP asymmetry may, in this way, have been determined by their preferred cognitive mode, rather than 'intelligence' *per se.*

Our results indicate that studies of EP lateral asymmetry must carefully control the cognitive mode of the subjects. As CALLAWAY [1973] has emphasized, the relation between EP and intelligence probably reflects the differences in subjects' ongoing cognitive processes rather than hard-wired differences in neurophysiological organization.

The lateralization of cognitive functions described above is characteristic of right-handed people. The cerebral lateralization of left-handed people is more complex. HÉCAEN and AJURIAGUERRA [1964] and HÉCAEN and SANGUET [1971], on the basis of an extensive review of the neurological literature and of their own clinical studies, concluded that left-handers show a greater cerebral ambi-laterality, not only for language, but also for gnostic and praxic functions. They distinguish between left-handedness which is familial and that which follows a perinatal injury to the left hemisphere. The familial type may or may not have reversed language lateralization.

These conclusions were generally confirmed by SATZ et al. [1967] in a study of a neurologically normal population. They used the dichotic listening test to assess language lateralization and carefully tested manual superiority rather than relying on the subjects' self-classification as to handedness.

Following the hypothesis of Orton that stuttering and dyslexia can be due to poorly established cerebral specialization, many studies have found high incidences of left-handers and ambidexterous people among these clinical groups. HÉCAEN and AJURIAGUERRA [1964] conclude that while no convincing direct relation has been demonstrated, 'disorders of laterality can play a part in a certain number of these cases'.

The nature of these 'disorders of laterality' is not clear. To our knowledge, there have been no attempts to quantitatively evaluate the interaction between the verbal-analytic and spatial-holistic cognitive systems in normal daily activities. Our opinion is that in many ordinary activities, normal people simply alternate between cognitive modes rather than integrating them. These modes complement each other but do not readily substitute for each other. Although it is possible to process complex spatial relationships in words, it would seem much more efficient to use visual-kinesthetic images. For example, consider what most people do when asked to describe a spiral staircase; they begin using words, but quickly fall back on gesturing with a finger.

Processing in the inappropriate cognitive system may not only be inefficient; it may actually interfere with processing in the appropriate system. This 'interference hypothesis' is supported by a study of left-handed subjects who were presumed to have bilateral language representation [LEVY, 1969]. Levy compared left-handed and right-handed subjects with equal WAIS verbal scores and found that the left handers had significantly lower performance scores, which she attributed to interference from the presumed ambilaterality of language. Her observation has been confirmed by MILLER [1971].

Similarly, in a group of patients in whom right-hemisphere language was demonstrated with carotid amytal, LANSDELL [1969] found a negative correlation between language ability and spatial performance scores. BROOKS [1970] presents additional support for the hypothesis of 'interhemispheric interference'. Reading a description of spatial relations interferes with the subsequent manipulation of those spatial relations. DENHYER and BARRETT [1971] demonstrated selective loss of spatial and verbal information in short-term memory by means of spatial and verbal interpolated tasks. LEVY has in fact suggested that verbal and nonverbal functions evolved in opposite hemispheres to reduce interference of one system with the other [LEVY, 1969].

This evidence of interference between the right and left cognitive modes provides a new kind of support for the hypothesis of Orton, that lack of cerebral lateral specialization plays a major role in dyslexia and stuttering. This hypothesis has continued to sustain interest, in spite of a lack of convincing direct evidence. Until recently, the only generally available index of cerebral lateralization was handedness, and people with little hand preference, or left handers who were 'switched', or those with mixed hand and eye preference were considered to be 'high risk'. The incidence of such people in clinical categories such as stuttering, dyslexia, and specific learning disability is usually found to be higher than in the normal population.

It is possible that our EEG method for studying lateralization of cognitive function, along with the dichotic listening test, can provide a much more direct and presumably more sensitive means for investigating disorders of laterality than measures based on hand, eye, or foot dominance.

Acknowledgment

This research was supported in part by an NIMH Career Development Award MH 28457, NIMH Grant MH 10307, and NIMH Interdisciplinary Training Program Grant MH 7082.

Dr. D. GALIN, Langley Porter Neuropsychiatric Institute, University of California, 401 Parnassus Avenue, *San Francisco, CA 94143* (USA). Tel. (415) 681 8080.

Language and Hemispheric Specialization in Man: Cerebral ERPs.
Prog. clin. Neurophysiol., vol. 3, Ed. J. E. DESMEDT, pp. 151–160 (Karger, Basel 1977)

Lateralization of Cortical Coupling during Complex Verbal and Spatial Behaviors

CHARLES D. YINGLING

Langley Porter Neuropsychiatric Institute, San Francisco, Calif.

The relative superiority of the left cerebral hemisphere for the production and comprehension of language has been recognized for over a hundred years, since it was first inferred from the greater vulnerability of language to left-sided brain damage. This view has been consistent with results obtained using a number of different experimental methodologies, including dichotic listening [KIMURA, 1964], unilateral carotid amytal injection [WADA and RASMUSSEN, 1960] and studies with split-brain patients [SPERRY, 1974]. Thus, it is surprising that electrophysiological attempts to obtain a correlate of this lateralization have met with relatively little success. A number of laboratories have attempted to demonstrate differential event-related potentials in the left and right hemispheres to linguistic or non-linguistic stimuli. Typically, the differences obtained have been small or non-existent, leading some workers to adopt a skeptical attitude toward the applicability of event-related potential measures to the assessment of laterality [FRIEDMAN et al., 1975a and this volume; GALAMBOS et al., 1975].

The method of averaged evoked potentials (EP) requires the presentation of large numbers of physically identical stimuli in order to obtain a satisfactory signal-to-noise ratio. This technique has led to a number of advances in our understanding of human sensory systems and is now proving itself to be of value in a number of clinical situations, as is amply documented elsewhere in these volumes. However, the necessity for repeated presentations of the same stimulus is an inherent drawback in the use of this method to study higher cognitive functions such as language. The repeated presentation of syllables such as 'pa' or 'ba' is not likely to engage the brain's cognitive linguistic analyzer in the same way as listening to an ordinary conversation. In fact, experiments with repeating tape loops have shown that even complex words and phrases undergo major changes in subjective intelligibility after far fewer repetitions than the number of stimuli presented in a typical evoked potential experiment [WARREN, 1968].

It would seem that a more reasonable search for EEG correlates of lateralized function would be to employ some means of analysis which does not place such rigid constraints on the experimental design, thus allowing subjects to perform complex verbal or spatial tasks which are more likely to produce strong lateralization. This requires a measure which is not time-locked to presentation of individual stimuli, but rather processes the ongoing EEG during some specified period of time. An obvious example is the use of frequency spectral analysis to obtain comparisons of power in various frequency bands between homologous areas of each hemisphere. This type of analysis has shown that alpha power is relatively suppressed over the left hemisphere during the performance of verbal tasks as compared to spatial tasks and vice versa [GALIN and ORNSTEIN, 1972; DOYLE *et al.*, 1974; GALIN and ELLIS, this volume].

Another approach is to compare the signals from different regions of the scalp and look for changes in their relationships as a function of various tasks. A simple method for obtaining such a comparison was described by CALLAWAY and HARRIS [1974]. Each channel is sampled every 4 msec and classified on the basis of polarity (+ or −) and whether its voltage is greater (↑) or smaller (↓) than the previous sample from the same channel. There are thus four possible categories (+ ↑, + ↓, − ↑, − ↓) and a 4×4 contingency table is tallied for each pair of channels for a fixed number of samples. The coupling between each pair of channels is computed from the contingency tables by comparing the actual cell frequencies with expected cell frequencies as calculated from marginal frequencies. The most familiar such statistic is χ^2, which requires many multiplications and rapidly becomes computationally cumbersome as the number of channels is increased.

An alternative method is the Shannon-Weaver information transmission (or uncertainty reduction) statistic Txy [SHANNON and WEAVER, 1949]. This measure can be considered analogous to χ^2 computed using log average frequencies and addition rather than multiplication. For a single channel (x) sampled over n points, uncertainty (entropy) is given by:

$$H_x = - \sum_i^n p(x_i) \log_2 p(x_i), \tag{1}$$

and for two channels (x and y) sampled simultaneously:

$$H_{xy} = - \sum_i \sum_j p(x_i, y_j) \log_2 p(x_i, y_j). \tag{2}$$

Reduction in uncertainty about x given y or vice versa is information transmission given by:

$$Txy = Hx + Hy - Hxy.$$

(3)

Since the sample size is fixed, and thus the range of possible values of p is predetermined, a table of $p \log_2 p$ values can be stored in the computer in advance and the entire computation involves only table look-up and fixed point arithmetic. As currently implemented on a NOVA 1220 minicomputer with 24K core memory, all possible couplings between 7 EEG channels can be computed on-line in real time; the 8th polygraph channel is used to monitor eye movements for artifact rejection purposes. If the same signal is paralleled into all channels, the theoretical maximum coupling is 2.0; in practice, this is reduced to 1.8–1.9 as a result of amplifier noise, digitizing error, etc., but the measure is still sensitive enough to discriminate reliably between different task conditions as will be shown below.

The original hypothesis of CALLAWAY and HARRIS [1974] suggested that if two areas of the brain are in active, functional communication with each other, the coupling between signals from these areas should increase. Thus, it was predicted that verbal tasks should increase the coupling between different areas of the left hemisphere relatively more than that between right hemisphere areas and vice versa for spatial tasks. The data have generally confirmed this prediction [CALLAWAY and HARRIS, 1974; YAGI et al., 1976].

Table I shows representative data from a single subject. EEG was recorded from F3, C3, P3, and 01, all referenced to the left earlobe, and from F4, C4, P4, and O2, referenced to the right ear. Coupling coefficients were computed for each pair of leads over the left hemisphere, and similarly for each right pair. The left/right ratio for each homologous pair is also shown. The subject alternated between writing from memory and performance of the WAIS block design task using only the right hand in each case. Three replications of each task were performed.

For this subject, the absolute magnitude of the coupling coefficients are almost always higher over the left hemisphere than the right, as reflected in the L/R ratios which are mostly greater than 1. However, these ratios are higher for the writing task than for the block design. As an example, for coupling between frontal and central leads, the mean L/R ratio for three repetitions is 1.39 for writing and 1.06 for block design. Similarly, for each of the other five pairs of leads, the L/R ratio is higher for writing. This is consistent with the hypothesis of relatively greater left hemisphere coupling during verbal tasks than during spatial tasks. When this same

Table I. Coupling coefficients and L/R coupling ratios during verbal and spatial tasks. Coupling is computed between each pair of leads over each hemisphere, for a total of twelve couplings and six L/R ratios for each task. Note the relative stability of the L/R ratios despite wide variations in absolute values of coupling coefficients. L/R ratios are consistently higher, for all lead combinations, during writing as opposed to block design. This indicates relatively higher coupling over the left hemisphere during verbal tasks, as predicted

	F3-C3	F4-C4	F3-P3	F4-P4	F3-O1	F4-O2	C3-P3	C4-P4	C3-O1	C4-O2	P3-O1	P4-O2
Writing I	0.562	0.358	0.365	0.186	0.181	0.082	0.910	0.671	0.391	0.210	0.697	0.517
L/R	1.57		1.96		2.21		1.36		1.86		1.35	
Writing II	0.338	0.303	0.226	0.161	0.119	0.176	0.885	0.654	0.370	0.188	0.641	0.464
L/R	1.12		1.41		1.56		1.35		1.97		1.38	
Writing III	0.518	0.351	0.325	0.181	0.153	0.077	0.865	0.617	0.373	0.183	0.689	0.507
L/R	1.48		1.80		1.99		1.40		2.04		1.36	
Writing (mean)	0.473	0.337	0.305	0.176	0.151	0.078	0.887	0.647	0.378	0.194	0.676	0.496
L/R (mean)	1.39		1.72		1.92		1.37		1.96		1.36	
Blocks I	0.524	0.470	0.342	0.242	0.162	0.099	0.881	0.682	0.340	0.222	0.645	0.534
L/R	1.12		1.41		1.63		1.29		1.53		1.21	
Blocks II	0.534	0.500	0.340	0.282	0.159	0.122	0.835	0.763	0.337	0.265	0.674	0.593
L/R	1.07		1.21		1.31		1.09		1.27		1.14	
Blocks III	0.410	0.417	0.274	0.214	0.146	0.089	0.888	0.637	0.395	0.202	0.722	0.528
L/R	0.98		1.28		1.65		1.39		1.96		1.37	
Blocks (mean)	0.489	0.462	0.319	0.246	0.156	0.103	0.868	0.694	0.357	0.230	0.680	0.552
L/R (mean)	1.06		1.30		1.30		1.26		1.59		1.24	

subject was tested 3 weeks later under the same conditions, the L/R coupling ratios for the frontal-central leads were 2.30 and 1.52 for writing and block design, respectively. This illustrates a point concerning the use of this measure to assess laterality which is probably applicable to many other situations as well. The absolute magnitude of measures within a given hemisphere is highly variable and likely to be influenced by a large number of factors other than the cognitive task. The *ratio* between hemispheric measures (be they coupling, alpha power, or whatever) is generally more stable, at least within a given experimental session. Even these ratios, however, may show wide variability from one session to another. The *shift in L/R ratios* when going from a verbal to a spatial task within a single experimental session has proved to be highly reliable, and the differences are consistently in the direction predicted. Thus, despite the enormous problems encountered in probing the intact brain for information relevant to different modes of complex cognition, there do appear to be clear and consistent effects that emerge if enough variables are held constant. Hopefully, repeated measures from the same subjects will allow the identification of many of these variables in order to better stabilize the performance of the measure.

There is an interesting situation in which exceptions to the general pattern are found, and analysis of this exception sheds light on the general problem of selection of appropriate measures to assess EEG lateralization. If a subject produces a high degree of alpha over the right hemisphere during a verbal task, the coupling measured over the right hemisphere may be high, reflecting the presence of alpha at all leads. (Generally, coupling coefficients are higher with eyes closed than with eyes open, reflecting this common-mode alpha effect.) Thus, coupling ratios in subjects with high alpha and strongly lateralized alpha suppression may be shifted in the opposite direction than that predicted.

This effect is demonstrated by the data of table II, which were obtained by analysis of data recorded on analog tape in an earlier study by ORNSTEIN and GALIN [1973, 1976].[1] The subjects in this study were chosen from two occupational groups characterized by primarily verbal or spatial activities (lawyers and potters). Each subject performed a series of tasks (copying text,

---

[1] Due to a multiplexing process in the FM recording system that was used some of the data channels on these tapes are slightly phase-shifted with respect to one another. While the coupling measure is relatively insensitive to phase (i.e. identical signals 90 or 180° out of phase still produce maximal coupling of 2.0), this caveat should be borne in mind. While the pattern of task-dependent coupling shifts was the same for the phase-shifted and non-phase-shifted channels, these studies should be repeated with data digitized on-line.

writing from memory, block design, mirror tracing) while EEG was recorded from P3, P4, T3, T4, C3, and C4, all referenced to vertex. For the purposes of the present comparison, the two verbal tasks have been combined as have the two spatial tasks. Right/left ratios of alpha power (9–11 Hz band) from the earlier study by ORNSTEIN and GALIN [1973, 1976] are compared with left/right coupling ratios obtained by analysis of coupling in the same EEG-recorded data; thus, for each method of analysis, the ratio should be higher for verbal than for spatial tasks.

Examination of table II shows that this prediction holds true for 11 of the 12 subjects when analyzed for alpha power and 11 of 12 when analyzed for coupling. Interestingly, the exception is not the same in each case. Subject P-8, who did not show the expected effect with coupling, was the subject with the largest alpha asymmetry. This subject produced large amounts of alpha, strongly lateralized according to task, even under the eyes-open conditions of the experiment. On the other hand, subject L-13 produced very low alpha power with right/left ratios in the opposite direction to that predicted; this subject showed coupling shifts in the expected direction. In general, there is a tendency for subjects with high alpha power and large task-dependent shifts in right/left alpha ratios to show poor shifts in cortical coupling and vice versa. This suggests that while alpha is the best indicator of lateralized hemispheric activation based on spectral power ratios [DOYLE et al., 1974], the relevant portion of the spectrum for obtaining task-dependent shifts in coupling lies outside the alpha range. This is perhaps not surprising in the light of the common view that alpha represents 'not operating' signals rather than information processing activity.

It is of some interest to note the difference in the two occupational groups. Lawyers tended to have higher left/right coupling ratios and higher right/left alpha ratios than did potters; this held for both verbal and spatial tasks and was also apparent during baseline 'attention to breathing' conditions. Similar differences between the two groups were noted in the original analysis by ORNSTEIN and GALIN [1973, 1976]. They further noted that the major difference between groups was in the task-related alpha shifts over the left hemisphere which were larger for the lawyers. It is tempting to speculate that these differences reflect a bias of the lawyers towards greater activation of their verbal cognitive mechanisms as compared with potters whose choice of occupation reflects a less verbal, more spatial bias. Whether these intriguing observations will prove to be reliable remains to be seen.

There are two possible approaches to the problem of alpha contamination which are currently under investigation. One method involves mathe-

Table II. Comparison of cortical coupling and alpha asymmetry in two occupational groups. Verbal tasks were copying text and writing from memory; spatial tasks were mirror tracing and WAIS block design. Both L/R coupling ratios and R/L alpha ratios are consistently higher for verbal than spatial tasks, with two exceptions (see text). Note tendency for lawyers to show higher ratios (i.e. greater left hemispheric activation) regardless of task. Data originally recorded by ORNSTEIN and GALIN [1973, 1976] and reprinted by permission of the authors

Subject No.	L/R ratios: cortical coupling			R/L ratios: alpha power (9–11 Hz)		
	verbal tasks	spatial tasks	baseline	verbal tasks	spatial tasks	baseline
P–4	0.99	0.93	0.99	1.19	0.69	0.85
P–5	0.99	0.96	0.90	1.05	0.82	1.48
P–8	0.93	1.29	1.01	1.28	0.69	0.83
P–9	1.33	1.20	1.07	0.71	0.54	0.64
P–10	1.01	0.81	0.94	1.46	1.20	1.15
P–15	1.68	1.08	0.84	0.98	0.81	0.67
P–18	1.05	0.98	0.84	1.08	1.06	1.64
Mean, 7 potters	1.14	1.04	0.94	1.11	0.83	1.04
L–2	1.55	1.44	1.21	1.12	0.94	1.40
L–5	1.24	0.98	0.96	1.61	1.27	1.51
L–11	1.31	0.92	1.05	1.21	1.15	1.29
L–13	1.05	0.92	0.95	0.91	1.00	1.33
L–16	1.32	0.94	0.88	1.02	0.94	1.14
Mean, 5 lawyers	1.29	1.04	1.01	1.17	1.06	1.33

matically partialling out the effect of posterior alpha from an occipital lead on coupling between more anterior pairs according to the formula:

$$Txy/z = Hxz + Hyz - Hz - Hxyz, \tag{4}$$

where Txy/z is information transmission between x and y when the effect of z is partialled out [BALI *et al.*, 1977]. In general, this approach has produced slight but non-significant improvements in discrimination between tasks and has not proven effective in overcoming large alpha effects such as those described above. This method is based on the assumption that alpha in more anterior leads represents activity which is volume-conducted from occipital generators. This assumption is questionable as different thalamocortical circuits apparently have the ability to function as independent facultative pacemakers [ANDERSEN and ANDERSSON, 1968; SKINNER and YINGLING, 1977].

A second approach is simply to filter alpha activity out of the EEG signal before the coupling is computed. This can be accomplished with analog notch filters, but more flexible results can be obtained with software-implemented digital filters [RABINER and GOLD, 1975]. Such an analysis is currently underway in collaboration with the EEG Systems Group, Langley Porter Neuropsychiatric Institute [YINGLING and GEVINS, in preparation]. Data is digitized on-line, and epochs containing slow (i.e. eye movement) or fast (i.e. EMG) artifact are deleted by the automated artifact rejection algorithms developed by the EEG Systems Group [GEVINS et al., 1975]. The data is subsequently analyzed both for power spectra and cortical coupling with a program which applies up to three user-defined, digital band-reject filters to the EEG signal prior to computation of coupling. The results of this study will be used to specify algorithms for implementation on the portable system.

In addition to the obvious theoretical interest of such an easily obtainable measure of lateralized function, there are numerous potential clinical applications. One such possibility is demonstrated in the study by BALI et al. [1977]. Following an observation by DAVIS and WADA [1974] that clicks increased coherence between averaged evoked potentials from two points over the left hemisphere relative to coherence between homologous points on the right hemisphere, whereas flashes had the reverse effect, BALI et al. measured coupling between parietal and frontal leads over each hemisphere while subjects were presented with trains of clicks or flashes at 3/sec. The data for 16 normal subjects are shown in table III. Note that all 16 subjects exhibited higher left/right ratios for clicks than for flashes, a result which is consistent with the suggestion that the left hemisphere, as a result of its specialization for analysis of language, is more strongly engaged by auditory stimuli than by visual input. It is also interesting to note that females, in general, showed higher left/right ratios than males, an intriguing observation in the light of the well-known fact that females score higher than males on the verbal portions of standardized intelligence tests whereas males are superior on the spatial tests.

In table IV are presented the results obtained in the BALI et al. study from a group of eleven adult male dyslexics recruited from a variety of remedial reading programs. As contrasted with 16 of 16 normals, only 1 of 11 dyslexics showed left/right ratios greater for clicks than for flashes. Further-

Table III. Comparison of coupling ratios during click or flash stimulation in males and females. L/R coupling ratios are higher during clicks for all subjects, indicating relatively greater left hemispheric activation by auditory stimulation. Note that ratios under both stimulus conditions are higher for females than for males. Data from BALI *et al.* [1977]

Subject No.	Sex	L/R (clicks)	L/R (flashes)	Clicks-flashes
1	M	1.185	1.087	+.098
2	M	0.969	0.855	+.114
3	M	0.889	0.756	+.133
4	M	1.147	1.026	+.121
5	M	0.994	0.974	+.020
6	M	0.989	0.964	+.025
7	M	0.997	0.877	+.120
8	M	0.989	0.978	+.011
9	M	0.843	0.829	+.014
10	M	0.985	0.900	+.085
11	F	1.358	1.135	+.223
12	F	1.218	1.186	+.032
13	F	1.325	1.286	+.039
14	F	1.489	1.442	+.047
15	F	1.094	0.872	+.222
16	F	1.336	1.077	+.259

Table IV. Comparison of coupling ratios during click or flash stimulation in dyslexic adults. Note that the higher ratio for clicks found in normal adults is not seen in these subjects with severe reading disabilities. Data from BALI *et al.* [1977]

Subject No.	L/R (clicks)	L/R (flashes)	Clicks-flashes
1	0.817	0.827	−0.01
2	0.983	1.058	−0.07
3	1.208	1.288	−0.08
4	1.238	1.322	−0.09
5	0.905	0.921	−0.02
6	0.906	0.902	+0.00
7	0.900	0.932	−0.03
8	0.990	0.995	−0.00
9	0.901	0.920	−0.02
10	0.814	0.845	−0.03
11	0.817	0.827	−0.01

more, the absolute magnitude of the change in coupling ratios between clicks and flashes is significantly smaller in the dyslexics than in the normal subjects. These data are compatible with the suggestion that reading difficulties may be related to a failure to develop normal patterns of lateralization. We have recently tested a small number of school-age children, including some with reading difficulties, and similar differences in lateralization to clicks and flashes between normal and poor readers have been found [HALLIDAY and YINGLING, unpublished observations]. The number of subjects tested to date is too small to draw firm conclusions at this point, but further investigation appears warranted.

In conclusion, it seems apparent that EEG measures which are consistent with current views of hemispheric specialization for linguistic and nonlinguistic processes can be reliably obtained if the analysis can be performed during complex, strongly lateralized tasks. The relative consistency between two such dissimilar measures as spectral power analysis and cortical coupling is encouraging, to say the least. The simplicity of the cortical coupling measure and its ease of implementation on a small, general purpose laboratory computer suggest that it may prove to be useful in a variety of research and clinical applications involving complementary specialization of the left and right hemispheres.

Acknowledgments

This work was supported by the National Institute of Mental Health Interdisciplinary Training Program MH–7082 and Office of Naval Research Contract N00014–75–C–0398. The author wishes to express his appreciation to DAVID GALIN and ROBERT ORNSTEIN for making available tape recorded data gathered in their laboratory and for valuable assistance in the analysis of the results, and to LEKH BALI, STEPHEN NAGHDI, ENOCH CALLAWAY, and ROY HALLIDAY for permission to quote unpublished results.

Dr. C.D. YINGLING, Langley Porter Neuropsychiatric Institute, University of California. 401 Parnassus Avenue, *San Francisco, CA 94143* (USA). Tel. (415) 681 80 80.

Language and Hemispheric Specialization in Man: Cerebral ERPs.
Prog. clin. Neurophysiol., vol. 3, Ed. J.E. Desmedt, pp. 161–171 (Karger, Basel 1977)

Quantitated EEG Correlates of Normal and Abnormal Interhemispheric Relations

Leonide Goldstein and Stevan R. Harnad

Department of Psychiatry, Rutgers Medical School, Piscataway, N.J.

It is now well established that there exist systematic relationships between brain wave morphology and brain functional states, and consequently behavior. Not only is this true in cases of organic dysfunction such as epilepsy, tumors, lesions and the like, but also in some psychiatric disorders: such states as anxiety, sedation, and others can be recognized by specific features of the organization of the electrical activity of the brain, even when the EEG is recorded transcranially as it is in man.

Until fairly recently, it was considered that 'ordinary' mental processes did not have detectable effects on the EEG. Reflecting back, it would appear that this was rather taken for granted than objectively demonstrated. As is usually the case in scientific investigation, no such relationships were known because no systematic efforts were made to detect them. This is just the more surprising when one considers that in the huge collection of EEG recordings examined by thousands of observers, no one had had the simplistic, yet potentially significant, idea to inquire as to the kind of thinking prevailing during the recordings in order to see whether any correlation existed between brain states and EEG features.

One of the most prominent gross EEG features is that of asymmetry between brain waves over the right (R) and left (L) hemispheres. Although often observable in the visual record, asymmetry was for a long time considered of no particular interest, perhaps because it was manifested in the amplitudes of the brain waves rather than in their frequency characteristics. Ever since the demonstration of its existence, the human EEG has tended to be conceptualized chiefly in terms of certain clusterings of frequencies, constituting the so-called 'basic rhythms'.

Human EEG studies began with the description of such a rhythm, the most easily recognizable, to which BERGER [1938] gave the name of 'alpha'. Later, other rhythms were described and in a general way related to brain functional states, such as concentration versus relaxation or waking versus sleep. However, since there were no visible differences between the rhythms on the R and L hemispheres, the amplitude asymmetry which did indeed exist was simply ignored as carrying no useful information. So pervasive was the over-emphasis on rhythms (and consequent underemphasis on amplitude) that WALTER stated in 1954, in unequivocal terms, 'the frequency of a rhythm is more significant than its amplitude'. As will be shown, recent research focusing on amplitude analysis, especially when it involves simultaneous recording from the R and L hemispheres, appears to shed some new light on the inter-relationships between behavioral states and brain electrical activity, particularly in terms of asymmetrical function. Furthermore, certain forms of psychopathology appear to be manifested in interhemispheric EEG amplitude relationships.

Methodology

A number of methods have been used for amplitude analysis. They are all based on the concept that the magnitude of the EEG signal is an expression of the electrical energy involved in its generation. It can be shown that the area subtended by the waves represents the square-root of the electrical power involved. Measurement of amplitude is accomplished by integrating the area of these waves (after first taking their absolute value in order to avoid negative and positive voltages). Technically, such procedures amount to an analog/digital conversion, which can be obtained off-line from computer programs or performed on-line with the use of electronic integrators [DROHOCKI, 1948].

This last method has been used in most of the studies to be reported here. In practice, the EEG is filtered prior to integration to remove very high and low frequencies. It is important to understand that although measurements are obtained on all waves, regardless of their time-durations (i.e. frequency characteristics), the methodology involves cumulation of the areas of successive waves until a particular predetermined threshold is reached. In the studies reported, EEG activity was predominantly between 3 and 20 Hz. At this point a pulse is delivered, followed immediately by a reset to zero of the integrators. The pulses are then counted during fixed 'epochs', which can very between 0.5 and 20 sec or longer, depending upon the experiment. The total number of pulses is thus proportional to the amplitude of the waves occurring during the epoch, irrespective of the number of waves involved. The epoch becomes the unit of measurement. Quantitative and statistical analyses are based on the distributions of the number of pulses per epoch. This presupposes that any condition being studied is to be documented by a sufficient number of epochs to permit valid estimates of the mean number of pulses per epoch, the between-epoch variance, and related parameters.

The main difficulty of such an approach is obviously the fragmentation of a continuous phenomenon into separate epochs. This is complicated by the fact that the duration of the epochs must be decided arbitrarily (on top of the previously mentioned arbitrary choice of the threshold for pulse delivery). In humans, it appears that with epochs of 10 sec and pulses delivered at the rate of 100 for 100 cumulated sine-wave oscillations of 50 μV peak-to-peak, one obtains reliable between-subject, between-state data. Furthermore, during a standard 10–min recording session, one obtains 60 values per electrode location with these parameters, which appears sufficient for valid estimations of both means and variances.

Among the advantages of the procedure, there is the fact that the digital counterpart of the analog signals can be 'edited', i.e. corrected for the occurrence of activity which is not primarily EEG, but due rather to movement or muscular activity artifacts. There is also the possibility, following simple transformations, to average between-subject data.

In laterality studies, activity from R/L pairs of homologous sites is analyzed in a number of ways. First, the time-course variations of amplitudes during the entire recording period are plotted. This produces graphs or 'chronograms' depicting the successive changes in each pair of electrodes. Secondly, the mean amplitudes and variances are computed for the R and L sides, as well as for their R/L ratios. Thirdly, one can calculate the R/L ratio for each individual epoch, as well as the overall mean ratio and the variance of the distribution of ratios. One can also compute the arithmetical difference between the successive individual ratios and their overall mean. When portrayed graphically, the latter permits determination of the proportion of epochs with relatively higher amplitude in one hemisphere or the other.

It is not always predictable which of the above-mentioned parameters will best reflect experimental effects; therefore, most, if not all of them, must be examined before any conclusion is reached. Different analytical procedures have differential sensitivity and, as will be illustrated below, certain factors affect the amplitudes more than the variances, and vice versa.

Another reason for diverse analyses is that changes in hemispheric EEG amplitudes, when present, are almost always of small magnitude and are made visible only under transformations whose optimality must be ascertained *a posteriori*. Since computer programs are available which rapidly perform all the previously mentioned calculations, the diversity of analytical procedures is not a serious hindrance to research.

Hemispheric EEG Relationships during Verbal and Non-verbal Mental Activity

In normal male and female adults, one finds that in occipital and temporal regions, the amplitude of the EEG on the left tends to be lower than on the right. According to SUGERMAN *et al.* [1973], in 58% of 67 normal volunteers, the R/L amplitude ratio was above 1.0. The R/L variance ratios were found to vary between 0.8 and 1.2. A possible reason for the higher proportion of subjects with R/L amplitude ratios above 1.0 may be that most subjects think verbally during the recording and that this is reflected by more activity over the left (language-dominant) hemisphere *(greater* activity is reflected in the EEG by *lower* amplitude).

In another study, 13 baseline recordings of 5 min duration each were obtained from right-handed college students, uninstructed as to what to think about during the recording session but asked merely to 'relax'. The time-course differences of the individual epoch R/L amplitude ratios from the overall mean ratio for the entire session reveal that the values were above 1.0 (indicating a greater activity in the L hemisphere) on the average during 66% of the recording time. Post-recording interviews revealed that in most cases subjects had experienced concern about the EEG procedures, apprehensions about the eventual findings, and other essentially verbal thoughts during the recording session.

A further reason for interpreting non-directed baseline recordings in terms of on-going verbal activity comes from experiments in which normal subjects were given instructions on what to think during the session. When the subjects were asked to think verbally (e.g. rehearse a speech in their minds or repeat mentally sentences remembered from books) and then non-verbally (e.g. think of shapes, forms, colors, musical tunes without lyrics), it was found that for most of them the EEG hemispheric laterality shifted from lower relative amplitude on the left during verbal activity to lower relative amplitude on the right during non-verbal activity.

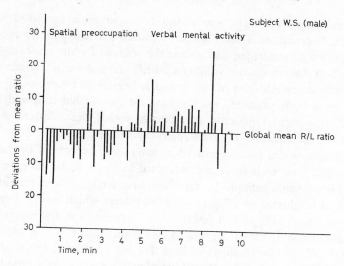

Fig. 1. Time-course distribution of temporal lobe R/L EEG amplitude ratios in a subject during undirected mental activity. As can be seen, during the period involving pre-occupation with spatial location of a noise source, the EEG amplitudes were relatively higher in the L hemisphere (indicating activation in the R), while during verbal processes, the relationships were reversed.

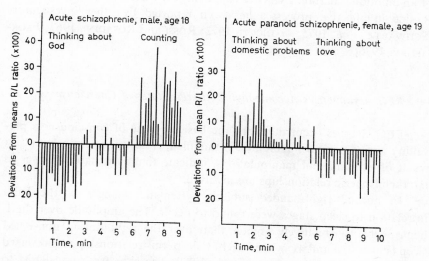

Fig. 2. Time-course distribution of temporal lobe R/L EEG amplitude ratios in schizophrenic patients during verbal and non-verbal mental activities. See text for detailed discription of the ongoing mental processes.

Similar changes have also been found to occur spontaneously, i.e. without specific instructions. Thus, a subject reported that during the recording session, he first concentrated on locating the source of white noise in the room, and then reflected on the validity of EEG data. As can be seen in figure 1, the relative amplitudes in the R and L hemisphere correspond to what would have been predicted on the basis of non-verbal followed by verbal activity. In mental patients, it was also found that clustering of deviations of the individual R/L amplitude ratios in the direction of one or the other hemisphere can be accounted for in terms of reports of ongoing mental processes. Two such cases are presented in figure 2. In the case of the male schizophrenic patient, he reported first feeling 'deep religious ecstasy'; this was followed by efforts not to fall asleep, which involved enumerating digits. The female patient described her experience as first involving contemplation of the financial aspects of her confinement in the Mental Health Center followed by 'feelings of love' for her boyfriend. In all three cases presented in figures 1 and 2, the electrical activity was clearly found to be predominantly greater in the L hemisphere during verbal thinking and in the R hemisphere during non-verbal activity.

It should be noted that in the case of the patients, the activity which brought about an increased involvement of the R hemisphere appeared to be of an emotional nature. Evidence suggesting an enhanced activation of the R side in affective processes has been reported by other investigators [SCHWARTZ et al., 1975; GAINOTTI, 1972; ROSSI and ROSADINI, 1967; FLOR-HENRY, 1976].

EEG Asymmetries Accompanying Altered States of Consciousness

EEG changes occurring when the baseline state of consciousness prevailing during normal wakefulness is modified, either spontaneously or as a result of experimental manipulations, indicate that systematic changes in interhemispheric relationships are involved.

In normal right-handed subjects systematic changes in the EEG in relation to sleep stages were found to occur. The amplitude over the L hemisphere was relatively greater than over the right during non-REM sleep (stages 2, 3 and 4); while in REM sleep, this relationship was reversed ($n = 7$; $p < 0.005$) [GOLDSTEIN et al., 1972]. The shift from non-REM to REM sleep is also accompanied by an overall reduction in amplitudes [AGNEW et al., 1967].

Possibly this indication of greater R hemisphere involvement in REM sleep is related to the visual imagery frequently reported to occur during dreaming. This would be consistent with the outcome of other studies reported above concerning directed non-verbal activity in waking subjects. A further indication in this direction is provided by the finding that drug-induced visual hallucinations also involve a reversal of predrug baseline EEG laterality relations [GOLDSTEIN and STOLTZFUS, 1973]. (Many similar changes have also been found to occur in rabbits and in cats [NELSEN *et al.*, 1977; WEBSTER, 1977].)

Spontaneously occurring and drug-induced changes along a continuum of arousal extending from anxiety to sedation are also accompanied by changes in interhemispheric relationships [GOLDSTEIN and STOLTZFUS, 1973]; but these tend to be less pronounced than changes during directed mental activity or sleep stages. Hyper-arousal induced by stimulant drugs such as amphetamine is accompanied by a decrease in the pre-drug amplitude asymmetry, i.e. as the amplitudes drop with the stimulant effect, they tend to approach equality over both hemispheres (table I).

Table I. Representative examples of the changes of mean amplitudes on the R and L hemispheres and of the R/L amplitude ratios produced by psychoactive drugs [GOLDSTEIN and STOLTZFUS, 1973]

Drug and total oral dose	Mean amplitude ± SD		R/L ratio
	R hemisphere	L hemisphere	
Pre-drug			
Post D-amphetamine	29.4 ± 4.5	26.3 ± 5.0	1.12 ± 0.08
15 mg (n = 10)	21.5 ± 2.7	20.8 ± 2.1	1.03 ± 0.05
Pre-drug			
Post D-675 25 mg[1]	76.1 ± 10.6	53.7 ± 7.5	1.37 ± 0.10
(n = 7)	16.5 ± 7.6	30.0 ± 5.0	0.52 ± 0.03
Pre-drug			
Post gluthetimide	53.4 ± 10.7	62.7 ± 10.4	0.85 ± 0.26
250 mg (n = 10)	56.7 ± 17.6	81.0 ± 19.3	0.70 ± 0.10
Pre-drug			
Post Charass[2]	52.3 ± 10.2	57.8 ± 8.4	0.94 ± 0.09
0.5 ml (n = 5)	51.9 ± 9.7	64.8 ± 12.0	0.81 ± 0.05

[1] Experimental analgesic which turned out to be a potent hallucinogenic.
[2] Alcoholic tincture of Cannabis containing 2 mg of delta-9 THC.

It appears that the above-mentioned reversal of amplitude relations due to hallucinogenic drugs, which are also stimulants [GOLDSTEIN *et al.*, 1963], is actually in some respects a magnified version of the stimulant effect with amplitudes now not only dropping and approximating but actually crossing over. An attempt will be made to interpret this effect below.

In contrast to the stimulant effect, the EEG laterality changes accompanying sedation (spontaneous, as well as drug-induced) and euphoria (induced by an alcoholic tincture of Cannabis) consist of a rise in overall amplitude and an increase in asymmetry (table I). The most pronounced manifestation of amplitude asymmetry observed to date occurs during masturbation-induced sexual orgasm, when the amplitudes over the R hemisphere are more than twice those over the L (n = 9, p<0.01) [COHEN *et al.*, 1976].

In all data cited thus far, irrespective of the magnitude of the overall amplitude or the amplitude asymmetry, there has always been a highly significant positive correlation between the amplitudes on the two sides (0.70<r<0.95) [GOLDSTEIN *et al.*, 1973]. One experimental condition has been found which totally abolishes this ubiquitous RL correlation. When a subject undergoes 1 h of perceptual deprivation ('sensory monotony'), the coefficient of RL correlation drops within 20–30 min to 0.03 [GOLDSTEIN *et al.*, 1973], indicating some sort of dissociation between the electrical activities on the two sides.

EEG Asymmetry in the Blind

The experiment on directed verbal and non-verbal mentation was performed on ten congenitally blind college students; four exhibited higher EEG amplitude over the left hemisphere during the undirected resting baseline while six had higher amplitude over the right. Handedness was not clearly expressed in these sightless subjects. Although all reported being 'right-handed', their responses to questions adapted to sightlessness indicated ambivalent hand use and it was found that some of them read Braille alternately with the right and the left hand.

During directed mental performance, it was found that, unlike normal subjects, both the verbal and the non-verbal conditions, as well as a condition in which they read Braille, all appeared to be mediated by the same hemisphere – the one opposite to the more active side during baseline (fig. 3). According to the subjects' reports, their activity during the baseline con-

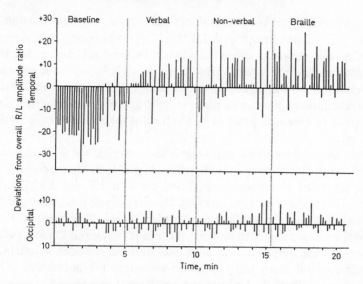

Fig. 3. Time-course deviations of the R/L EEG amplitude ratios in a congenitally blind subject during four successive sequences: baseline without specific instructions, requested verbal mental activity, requested non-verbal mental activity (music tune), and reading Braille. On top, temporal lobes (T3–T4); on the bottom, occipital lobes (O_3–O_2). It will be noticed that in that subject the activity in the occipital was equitable to random noise.

sisted chiefly of attending to proprioceptive and kinesthetic information. This suggests that, unlike in normal sighted subjects whose hemispheres differ in terms of verbal versus visuospatial and auditory function, in congenitally blind subjects both these functions are performed by one hemisphere while the other is specialized for somatosensory information processing. Also noteworthy is the fact that unlike non-blind subjects, the activity in the occipital lobe was not only unrelated to mental activity but, in fact, clearly indicative of random noise (fig. 3). This was found to be true of 6 of the 10 subjects.

EEG Asymmetry in Psychopathology

In the studies described so far, the parameter which best reflected interhemispheric relations was the mean amplitude ratio. In psychopathology the variance ratio turns out to be the more sensitive measure. As mentioned earlier, in normal subjects the R/L variance ratio during undirected baseline recordings ranges between 0.8 and 1.2. In 17 right-handed acute schizophrenics, this ratio ranged between 0.2 and 0.9. Of these schizophrenics, the paranoid patients had the lowest variance ratios while hebephrenic patients were clustered toward the higher end of the range [ROCHFORD et al., 1976]. Symptomatic improvement due to pharmacotherapy with neuroleptic drugs is accompanied in paranoid patients by a return of the variance ratio to the normal range.

In depressive illness (both psychotic and neurotic), the R/L variance ratio differs from the normal in a direction opposite to that observed in paranoids. Reported elevations have varied between 1.7 and 3.9 [D'ELIA and PERRIS, 1973; ROCHFORD et al., 1976]. Here too, symptomatic improvement due to either electroconvulsive or pharmacotherapy is accompanied by a return of the variance ratio to its normal range [D'ELIA and PERRIS, 1973].

The above R/L amplitude variance data as well as evidence obtained from power spectral analysis by other investigators [FLOR-HENRY, 1976] suggest an abnormality in the L (language-dominant) hemisphere in paranoid patients and in the R hemisphere (which is possibly involved in affective states) in depressed patients. This is consistent with the symptomatic behavior of the paranoid, which is hyper-verbal and hyper-rational while the depressed patient is hypo-verbal and dysphoric.

Discussion

A simple interpretation of the above data would be that in the normal waking state the hemispheres function as a strongly yoked system, but one which permits a shift in relative emphasis toward one side or the other in accordance with the specialized cognitive demands of the immediate situation. If this capability to shift is reduced, activity may be biased toward one side or the other in a non-optimal way and behavioral impairment with diminished information-processing efficiency, as well as anxiety, confusion, and stereotyped manifestations, may result. In the psychiatric conditions discussed, the hemispheric bias appears to have a 'positive feedback' character with the patient maladaptively 'stuck' in one hemispheric mode or the other, verbal for paranoids and non-verbal/dysphoric for depressives.

The relation of relative hemispheric activity to the continuum of arousal states is less clear-cut, but maximal lability of the capacity to shift between the hemispheres appears to prevail in the normal arousal range while hyper-

and hypo-arousal induce opposite hemispheric biases. It is in the normal arousal range that subjects can with relative ease voluntarily shift from verbal to non-verbal processing modes in accordance with directions, or spontaneously during 'routine' thinking processes.

The shifts in hemisphere emphasis occurring during sleep stages would appear to be normal endogenous asymmetries, perhaps related to the irrational, visual nature of dreaming, during which R hemispheric activity predominates. Lateralized EEG activity during sexual orgasm may be related to the hedonic intensity characteristic of this state.

In blind subjects, the R/L 'division of labor' [HARNAD, 1973] does not seem to be along the usual lines since one hemisphere is more active during both verbal and non-verbal processing while the other predominates during somatosensory perception. The only exception to the normally observed strong coupling between the electrical activities of both hemispheres, irrespective of relative R/L bias, occurred during perceptual deprivation, a condition which is thought to involve maximal overall impairment of information-processing capacity.

It thus appears that for a variety of normal as well as abnormal states, and/or conditions, amplitude analysis of R/L relations reveals features which may have a bearing on the nature of brain functional mechanisms. It remains to be determined why some of these effects are reflected in the R/L ratio of mean amplitudes while others are expressed in the R/L variance ratios, i.e. to ascertain to what feature of cortical processing each parameter is responsive.

Prof. L. GOLDSTEIN, Department of Psychiatry, CMDNJ-Rutgers Medical School, Piscataway, NJ 08854 (USA). Tel. (201) 564 4416.

Language and Hemispheric Specialization in Man: Cerebral ERPs.
Prog. clin. Neurophysiol., vol. 3, Ed. J.E. DESMEDT, pp. 172–187 (Karger, Basel 1977)

Search for Right Hemisphere Asymmetries in Event-Related Potentials to Somatosensory Cueing Signals[1]

JOHN E. DESMEDT and DONALD ROBERTSON

Brain Research Unit, University of Brussels, Brussels

Studies of cerebral event-related potentials (ERP) in man have been concerned for many years with components recorded at the midline vertex, like the Contingent Negative Variation (CNV) [WALTER et al., 1964; McCALLUM and KNOTT, 1973, 1976] and the P300 components [SUTTON et al., 1965, 1967; DESMEDT et al., 1965; DEBECKER and DESMEDT, 1966; DONCHIN and COHEN, 1967; VAUGHAN and RITTER, 1970; TUETING et al., 1971; HILLYARD et al., 1971, 1973; HILLYARD and PICTON, 1978]. While the possibility has been entertained at one time that P300 might not represent an independent phenomenon and could result from the resolution of the slow negative shift of the CNV [KARLIN, 1970], this issue has now been clarified and the P300 clearly dissociated from CNV [DONCHIN et al., 1975, and this volume; SYNDULKO and LINDSLEY, 1977; DESMEDT and DEBECKER, 1978].

Current interpretations suggest that the P300 is not related to the physical characteristics of the evoking stimulus, but occurs when a target stimulus delivers task-relevant information enabling the subject to make a perceptual decision [SUTTON et al., 1967; DONCHIN and COHEN, 1967; HILLYARD and PICTON, 1978; DESMEDT and ROBERTSON, 1977]. Without taking the extreme skeptical position of DONALD [1978] who is expressing the opinion that transient ERP components such as P300 might not index any aspect of stimulus processing, one should realize the need for better definition of the various ERP components which can be manipulated through psychological parameters. We are now only at the beginning of the analysis of the electrocortical correlates of brain processing and the available data are scarce and

[1] This research has been supported by grants from the Fonds de la Recherche Scientifique Médicale and from the Fonds National de la Recherche Scientifique of Belgium.

have by far not explored all the existing potentialities. Under such conditions, negative or discouraged statements [GALAMBOS *et al.*, 1975] may perhaps be helpful to encourage more lucid self criticism on the part of the investigators concerned, but such statements should not be taken to imply that ERP research might be hopeless.

Only in recent years have adequate methodologies been instrumented and the recent report on publication criteria of ERP data [DONCHIN *et al.*, 1977] should prove helpful in this respect. Many of the earlier experiments presented technical inadequacies due, for example, to lack of proper control of the eye movements and blink artifacts, to insufficient randomization of the trials and to inadequate recording system bandpass or control for amplifier blockings [cf. DESMEDT, 1977]. The experimental designs themselves may be in need of a thorough and lucid critical discussion. Many studies have been using rather artificial stimuli such as photic flashes which are now recognized as entirely inadequate as visual stimuli [ARDEN *et al.*, 1977; SPEKREIJSE *et al.*, 1977], acoustic clicks or electrical shocks on nerves. The sensory signals elicited by such stimuli would indeed appear unlikely to require elaborate cerebral processing for allowing the task to be carried out.

One difficulty in ERP research on higher brain functions is that the extraction of ERP requires the electronic averaging of many trial samples and the repetition of the same stimuli which can rapidly lose their attention-mobilizing capacities in the course of the runs. Several techniques have been used recently to attempt and bypass these constraints. For example one can present to the subject more or less continuous natural tasks which will load the brain processing sufficiently to engage special processors, and then deliver during this task brief probe stimuli in the same sensory modality. The idea is that the ERPs to the probe stimuli may be different according to the processing set of the brain [GALIN and ELLIS, 1975, and this volume].

Another procedure is to present different stimuli in random sequences at high rates whereby the cognitive overload elicited by the forced paced task will constrain the subject to use selective processing strategies [DEBECKER and DESMEDT, 1971; HILLYARD *et al.*, 1973; SCHWENT and HILLYARD, 1975; SCHWENT *et al.*, 1976; DESMEDT and ROBERTSON, 1977]. In spite of the interest of these strategies involving simple stimuli, it remains that further elaboration of the ERP approach to higher brain processing will have to rely on the design of more adequate and more natural stimulus configurations which would by their intrinsic structure engage special processors.

The present paper reviews recent data searching for lateral asymmetries of ERP components in several experimental paradigms. We have been interested in exploring the somatosensory modality for which the earlier cortical events are more easily recorded [DESMEDT, 1971; CRACCO, 1972; DESMEDT et al., 1974, 1976] and appear essentially on the side contralateral to the stimulated hand. It was thus thought that later components associated with appropriate psychological tasks might well be lateralized. Another reason for this approach is that some of the higher somatosensory functions appear to involve the right hemisphere in right-handed individuals, as suggested by clinical observations on patients with unilateral brain lesions [DIDE, 1938, ALAJOUANINE, 1948; TEUBER and WEINSTEIN, 1954; ETTLINGER et al., 1957; PIERCY et al., 1960; TEUBER, 1962, 1963; HÉCAEN, 1962; SEMMES et al., 1963] or on patients with a surgical transection of the cerebral commissures [BOGEN, 1969; BOGEN and BOGEN, 1969; SPERRY, 1974].

Methods

The experiments were carried out on normal unpaid volunteers of both sexes. The subjects (scientific staff, medical students, the authors) were highly motivated to perform the decision tasks correctly; a number of proposed subjects were discarded because of inadequate motivation in the psychological tasks, excess alpha in the electroencephalogram (EEG) or inability to properly relax so as to minimize eyeblinks and muscle potentials. The subjects sat comfortably in an armchair, in a sound-proofed air-conditioned electrically shielded room. About a dozen runs with alternating task conditions were carried out and the experiment was terminated before the performance was deteriorated by fatigue.

The brain potentials were recorded with stainless steel sterile needles inserted into the scalp to the same length of 6 mm. The recording system bandpass extended from 3 kHz to 0.05 Hz (or 3 sec time constant) and this wide fidelity was necessary to resolve both the slow components and the early components of the somatosensory evoked potentials (SEP) [cf. DESMEDT et al., 1974; DESMEDT, 1977]. The recordings from the parietal (3 cm behind the C positions of the international 10–20 system), central and frontal positions were as a rule referred to the earlobe on the same side. The vertex (Cz) derivation was referred to the left earlobe. The potentials of the electrooculogram (EOG) was recorded with Beckman cup electrodes fixed to the orbital ridge above and below the right eye. The EOG was averaged over the same trials as those retained for the brain potentials, as a check against eye movement interference. All potentials were recorded on FM tape at 15 ips. The data were then edited off line and averaged by a Nicolet 1074 digital computer (4,096 words of 9 bits). The resolution was 4.1 points/msec [cf. DONCHIN et al., 1977a].

Two series of experiments were performed with a four-finger paradigm in which electrical square pulses of 0.2 msec duration were delivered in a random sequence to the second and third fingers of either hand. The subjective threshold for each stimulus was determined by the method of limits; the intensity measured by a current probe was then set at 1 or 2 mA above the threshold. This provided a clear but rather small subjective sensation and the subject could concentrate his attention to the stimuli delivered to one of the fingers. The differentiation of target stimuli to the designated finger from the non-target stimuli to the adjacent finger was however rather difficult in the experiments at faster rates. The stimuli were frequently checked for intensity and to ensure that they remained subjectively equivalent on the four fingers throughout the session.

The stimuli were delivered in random sequence on the four fingers, and the intervals between any pair of consecutive stimuli also varied at random between set limits. The random intervals were provided by a beta emitter [CARMELIET et al., 1971]. In the slow four-finger paradigm, the mean random interval between consecutive stimuli was 1.5 sec for all fingers and 6 sec for any one finger. In the fast four finger paradigm, the mean random rate between consecutive stimuli was 400 msec for all fingers and 1.6 sec for any one finger. Further details can be found elsewhere [DESMEDT et al., 1977; DESMEDT and ROBERTSON, 1977; DESMEDT, 1977c].

In a third series of experiments an attempt was made to design a paradigm involving more natural tactile stimuli which would be more meaningful and more likely to engage spatial orientation processing mechanisms. The subject was to palpate with the index finger the top of a vertical plexiglass rod presenting a circular ridge interrupted by a gap of 1.5 mm. The rod was mounted on the shaft of a step motor that was interfaced to the electronic program. The orientation of the ridge gap was changed between trials to any one of four positions at 90° from each other. The finger was lifted by an electromechanical device between trials. The trial was initiated as the finger was dropped onto the ridge. A few seconds after the trial the finger was lifted automatically in preparation for the next trial. Control conditions were provided by a similar active touch task carried out on a smooth surface.

Slow Four-Finger Paradigm

The intramodality selective attention task involving a designated finger was rather easy to carry out, even when using weak stimuli. The subjects counted the target stimuli mentally during each run and made few mistakes. The situation was straightforward since there were no warning signals, the intervals varied at random and any expectancy effect in the series was excluded. The subjects made no motor response to avoid the cerebral potentials associated with movements [cf. DEECKE and KORNHUBER, 1977]. A large P300 component appeared consistently in the SEPs to the target finger stimuli. The P300 was small or virtually absent in the SEPs to physically identical stimuli delivered in alternate runs when the subject attended a designated finger in the opposite hand (fig. 1B). We confirmed RITTER et al. [1972] who indicated that P300 was maximal over the parietal regions. However the pertinent derivations for the present work were not from the midline vertex but from the two parietal areas. In the derivation contralateral to the target finger stimulus, early components including the positive P45 were recorded: these were not modified by the task conditions and indeed super-

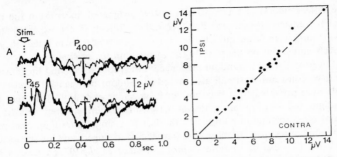

Fig. 1. Four-finger paradigm at slow random rate and the distribution of P400. *A, B* Cerebral SEPs evoked by electrical stimuli of 2.8 mA (0.8 mA above subjective threshold) to the second finger of the right hand. *A* Ipsilateral parietal scalp derivation (3 cm behind C4). *B* Contralateral parietal scalp derivation (3 cm behind C3). Each active electrode is referred to a separate reference electrode inserted into the earlobe on the same side. A random sequence of four different stimuli of equivalent intensity is delivered to four fingers, namely the second and third fingers of either hand. The mean intervals between consecutive stimuli on any finger is 1.5 sec (or 6.0 sec for the mean interval on the same finger). The thicker traces represent SEPs to the attended target stimuli. The thinner traces represent SEPs to identical stimuli to the same finger, but recorded in other runs when the subject attends stimuli in the opposite hand. A large P400 SEP component is elicited by the target stimuli only. The earlier SEP components of latencies below 200 msec are not significantly affected by the task. The contralateral SEP (B) presents an early component P45 which is not seen in the ipsilateral SEP (A). Negativity of the active electrode produces an upward deflection in all traces. *C* Pooled data for a total of 20 experiments on 12 subjects. Each point corresponds to a different experiment in which the peak voltage (μV) of P400 to target stimuli is plotted for the simultaneously recorded contralateral (abscissa) and ipsilateral (ordinate) SEPs. Target stimuli were delivered to fingers in either right or left hands. There is no significant deviation from proportionality of P400 on both sides. From DESMEDT *et al.* [1977].

impose quite consistently in figure 1B. The early components were absent at the ipsilateral derivation in figure 1A where a large P300 was recorded for the target finger stimuli only.

It has been usual to talk about P300 for designating the late positive component elicited in ERPs by target stimuli and which has a peak at about 250–450 msec. However in any given publication it may be preferable to consider the actual peak latency of the components recorded and to include this mean latency for designating the phenomenon [cf. DONCHIN *et al.*, 1977a]. Therefore we use P400 which is close to the mean peak latencies recorded in these series of experiments (376 and 396 msec) [DESMEDT *et al.*, 1977; DESMEDT and ROBERTSON, 1977].

Figure 1C presents pooled data about the lateral distribution of P400 voltages. The amplitudes were measured from the prestimulus baseline for the simultaneously recorded contralateral and ipsilateral SEPs. The data obviously fit a 45° line thus showing that P400 is symmetrically distributed at the two parietal areas in these experiments.

Bisensory Paradigm

When the stimuli are delivered at such rather slow rates to two fingers and intermixed with acoustic clicks, a P300 is also recorded when the subject attends the target finger stimuli (fig. 2). The baseline records are provided by SEPs to identical stimuli recorded in runs when the subject attends the clicks. A difference here is that in addition, a negative component with a peak at about 150 msec, the N150, is recorded to the target shocks. The early components of the SEP, the N20 and the P45 are not affected by the task (fig. 2B). The N150 is increased on both sides, although the effect is generally larger at the contralateral derivations (see fig. 4A). The P400 presents about the same voltage at the parietal derivation on the two sides. Figure 2 also illustrates two controls, namely the EOG which is flat and indicates that absence of eye movements contamination in the averaged samples (A), and the add-sub average for the contralateral samples (D). In the add-sub the successive samples are alternatively added and substracted [SCHIMMEL, 1967] thereby eliminating the ERP consistently associated in time with the epoch and presenting the nonresponse background not so coherently related to the stimulus. The add-sub provides a check on possible contributions from the background to the averaged ERP. The add-sub is virtually flat in this example and shows that no artefactual deflections contributed to the waveform of figure 2B.

Fast Four-Finger Paradigm

An interesting finding is that no task-related increase of N150 is recorded in the four-finger paradigm carried out at the slow rate of 40/min or 0.66/sec (fig. 1A, B). However, this is elicited when the mean random rate is increased as in figure 3. Only the contralateral derivations are illustrated, for either SEPs to target stimuli (A, C) or to nontarget stimuli delivered to the finger adjacent to the one attended in the same hand (B, D). The superimposed

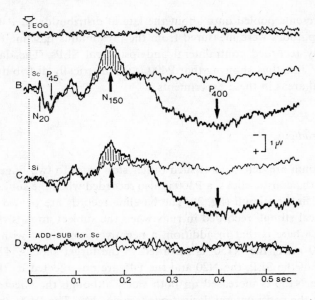

Fig. 2. Cognitive SEP components N150 and P400 at right and left parietal derivations. *A* Vertical EOG averaged for the selected trials used to produce the SEPs in B and C; the EOGs document the absence of any remaining eye movement artifacts in these samples. The SEPs were elicited by electrical stimuli of 3.5 mA to the third finger of the right hand in runs when the subject attends that finger (thicker traces) or in other runs when he attends acoustic clicks delivered through earphones (thinner traces). *B* Contralateral SEP recorded from the parietal scalp derivation, 3 cm behind C3. *C* Ipsilateral SEP recorded from the symmetrical derivation, 3 cm behind C4. *D Add-sub* averages of the SEP samples which are presented in B as *add* averages; the add-sub average checks for possible contribution of nonresponse EEG background to the averaged SEP response. The increase of N150 to target stimuli (cross-hatched area) is somewhat larger contralaterally (B). The P400 estimated from prestimulus baseline is virtually equal on both sides.

thinner traces provide a control condition, namely the SEPs to identical stimuli when the subject attends a designated finger in the opposite hand.

The early SEP components N20 and P45 are not affected by the task. An increase of N150 by the task (cross-hatched area) occurs earlier and is larger for the faster rate of 150/min or 2.5/sec (fig. 3C, D). The effect, though smaller, is present also in A and D for the intermediate rate of 75/min or 1.25/sec. Notice that the N150 increase is recorded both for the target stimuli (A and C) and for the nontarget stimuli (B and D) of the same attended hand.

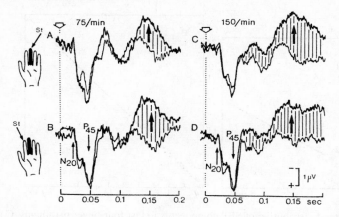

Fig. 3. Effect of mean random rate of stimulation in the four-finger paradigm on the earlier SEP components. All traces were recorded in different runs of the same experiment. Electrical stimuli of 3 mA were delivered (arrows in the figurines) either to the third finger (A,C) or to the second finger (B,D) of the left hand. The figurines present in black the fingers which are designated as targets to be counted by the subject in any given run. The thicker traces are SEPs to target stimuli (A,C) or to nontarget stimuli to adjacent fingers (B,D) when the subject attends the same (left) hand. The thinner traces are the SEPs to identical stimuli, but in runs where the subject attends target fingers in the opposite hand. The early N20 and P45 components are not significantly changed by the tash. The later negative N140 component is increased for both target and nontarget stimuli in the attended hand (cross-hatched area). The increase is larger and occurs at an earlier latency after the stimuli in the experiment carried out at the higher mean random rate of 150/min (C,D).

The four-finger paradigm can thus easily be carried out at different random mean rates and demonstrates that the forced paced conditions with the same simple stimuli can determine the emergence of an additional cognitive component which precedes the P400 phenomenon.

Figure 4 presents pooled data about the lateral distribution of the cognitive components in the fast four finger paradigm (150/min mean random rate). The voltage of P400 is again symmetrical at the left and right parietal areas (B). By contrast, the N150 is definitely larger on the contralateral side (A). In this graph we also plotted a P100 component which appeared in some of the subjects and was increased for target stimuli only; the P100 cognitive component also disclosed a definite lateralization to the contralateral parietal area. Further details on these recent results can be found elsewhere [DESMEDT and ROBERTSON, 1977]. It is important to stress that the preponderance of N150 and P100 occurred on the side contralateral to

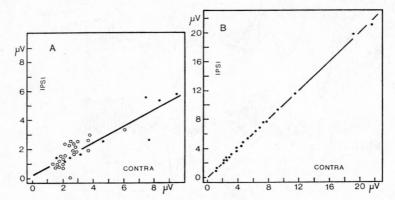

Fig. 4. Lateral distribution of SEP components in fast four-finger paradigms. Comparison of the peak voltages of the cognitive components N140 (circles in A), P100 (dots in A) and P400 (dots in B) at symmetrical parietal derivations referred to the earlobe on the same side. The voltages at the contralateral (abscissa) and ipsilateral (ordinate) parietal locations are compared. The voltages in μV are estimated from the prestimulus baseline. The SEP components N100 and N140 are larger contralaterally to the attended hand, no matter whether the target stimuli are delivered to the right or left hand ($y = 0.23 + 0.55\,x$; $r^2 = 0.79$). The P400 components are equal on both sides ($y = 0.04 + 0.99\,x$; $r^2 = 0.99$). From DESMEDT and ROBERTSON [1977].

the target finger stimuli considered in the corresponding runs: these enhancements did not selectively involve either the left or the right hemisphere as such. One interesting finding of these experiments is that P100 might index target signals whereas N150 might index both target and nontarget signals arriving over the attended channel [DESMEDT and ROBERTSON, 1977].

The lateral electrogenesis elicited by the selective attention task carried out at fast rates must be thus related to a special involvement of the posterior parietal associative areas on the side where the somatosensory input arrives and is first organized in the primary projection areas. This lateralization is however not absolute (fig. 4A) and is not related to any unique specialization of the hemispheres.

Active Touch Paradigm

Experiments along an entirely different line were initiated to find out whether lateralized ERP components could be elicited in leisurely paced tasks which would involve natural and more meaningful somatosensory stimulus conditions. Psychological testing with map reading or with identification of objects by manipulation have been used in studies of spatial orientation and they had suggested the significance of the right hemisphere for such behavior [cf. TEUBER, 1963; SEMMES et al., 1963; SEMMES, 1965; FREDERIKS, 1969; MILNER, 1971; FONTENOT and BENTON, 1971; NEBES, 1971, 1972, 1974]. Complex tasks of that type could be used while investigating the spontaneous EEG changes in the two hemispheres with methods such as spectral analysis or cortical coupling [cf. GALIN and ELLIS, DAVIS and WADA, YINGLING, all this volume].

For investigating transient ERPs another task had to be designed which would be carried out in a rather standard repeatable manner, and during a short time so as to allow coherent averaging. After several attempts we found that orientation in extrapersonal space could be instrumented in relation to an active touch task [cf. GIBSON, 1962] in which the subject had to palpate a plexiglass ridge in order to find the position of a small gap in the ridge. The timing of the task was determined by using an electromechanical device that lifted the finger from the board between trials and dropped the finger onto the ridge to initiate each trial of the run.

Active touch is a complex behavior for acquiring specific somatosensory information. It involves tactile scanning and small movements of the finger. The motor commands for such movements are no doubt associated with cerebral activities such as indexed by the readiness and motor potentials [DEECKE et al., 1969, 1976, 1977; KORNHUBER, 1974; DEECKE and KORNHUBER, 1977]. The readiness potential is largely symmetrical while the motor potential in the central area is contralateral to the moving hand. It is also known that the cerebral motor outflow to the motoneurons of the distal limb muscles is entirely crossed [BRINKMAN and KUYPERS, 1972; KUYPERS, 1973]. Thus the lateralization of the motor potentials to the opposite side of the active limb provides a useful criterion when searching for right hemisphere ERP activities in relation to active touch: when using either the right or the left hand for the task, the motor potential should cross sides whereas any ERP components related to higher somatosensory processing should be expected to occur or predominate on the right side.

Figure 5 illustrates a successful experiment along this line. A right-handed subject carried the active touch task with the right index finger which was dropped onto the ridge at the time indicated by the arrow and vertical dotted line. The thicker traces correspond to averages in runs carried out with the ridge as described. The thinner traces correspond to averages in other runs carried out similarly as controls, but with the finger dropped

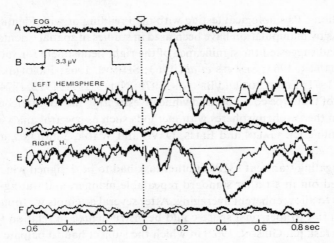

Fig. 5. Right parietal increase in ERPs during active touch in a right-handed subject. At the time indicated by the hollow arrow and the vertical dots, the right index finger is dropped by a mechanical device onto a circular ridge on a perspex rod. Palpation allows the subject to identify the orientation with respect to his own body of a narrow groove in the ridge (thicker traces). The thinner traces correspond to other runs carried out in the same way but with the ridge replaced by a smooth perspex surface. *A* Vertical EOG averaged for the trials used to produce the ERPs; these EOGs document the absence of any remaining eye movement artifact in the samples. *B* Calibration step function of 3.3 μV. *C–F* Cerebral potentials simultaneously recorded from the left (C, D) and right E, F) parietal regions, 3 cm behind the central locations C3 and C4; the active electrodes are referred to the earlobe on the same side. *C, E Add* averages displaying the ERPs. *D, F Add-sub* averages of the same EEG samples which checks that the nonresponse EEG background does not contribute to the waveforms of the ERPs. Notice that the left parietal recording in C presents large early components, namely a positive P50 and a negative N150, which probably represent the early contralateral electrogenesis evoked by the tactile contact of the right index finger on the perspex. For latencies greater than about 250 msec, the ERPs over the left hemisphere are not much different for ridge or smooth surface, whereas the ERPs over the right hemisphere (in E) disclose a large positive component for ridge palpation (thicker trace).

onto a smooth plexiglass surface which carried no spatial information to the subject. The EOG averages in figure 5A indicate that no eye movement or blink contamination remained in the trials selected for averaging. In the derivation of the left parietal cortex (fig. 5C), the finger contact with the plexiglass elicits early components and it is interesting that a large negative component of about 180 msec peak latency (probably equivalent to a N150) is enhanced when touching the ridge as compared to the smooth surface.

This may be compared to the contralateral N150 effect described in the preceding section and in DESMEDT and ROBERTSON [1977] for simple finger shocks in the fast four-finger paradigm. However the subsequent ERP components on the left side disclose not much difference between the ridge and smooth surface runs. In figure 5D, the averaged add-sub control of the same samples shows that no change in nonresponse background activity could account for the ERP features just considered.

At the right parietal cortex (fig. 5E) the 'N150' negativity is not conspicuous which is not surprising in view of our finding that N150 is lateralized to the side opposite to the hand receiving somatosensory input in the experiments reported in the preceding section (fig. 4A). This example shows marked lateralization of N150. The main finding is however an important and prolonged positive electrogenesis recorded over the right side for the runs with the ridge; this effect is much smaller in the runs with smooth surface. The add-sub control (fig. 5F) again excludes that variations in nonresponse background might account for these features.

The experiment of figure 5 shows that an active touch paradigm involving spatial orientation can elicit a lateralized positive electrogenesis over the right hemisphere in the right-handed subject. We could confirm this finding in several other subjects. Moreover when the active touch task is carried out by the left hand, the positive electrogenesis remains on the right side of the brain [DESMEDT, 1977d].

Discussion

As the extensive clinical evidence for hemispheric specialization is considered [ETTLINGER et al., 1957; PIERCY et al., 1960; SEMMES et al., 1963; BOGEN, 1969; SPERRY, 1974] and also the anatomical evidence showing consistent differences in the parietotemporal areas on the right and left sides [GESCHWIND and LEVITSKY, 1968; WITELSON and PALLIE, 1973; WADA et al., 1975; RUBENS et al., 1976; CHI et al., 1977], it becomes obvious that cerebral ERPs should, if feasible, be developed into a major tool for the further investigation of these phenomena. The ERP approach is important, not merely to confirm the existing data, but to expand such results in normal individuals with intact brain. The major problem however is to enquire whether appropriate tasks and experimental conditions can be designed to elicit ERP components which are lateralized and which can index these special processing functions. Much of the existing evidence is still inferential or inconsistent [cf. DONCHIN et al., this volume], but encouraging results are now emerging, as shown in various chapters of this book.

It is perhaps important to systemize a number of features of ERPs in relation to the lateralization issues. It must indeed be realized that lateralization of ERP components can have very different significance, as appears in the following classification which is proposed as a first approximation based on our data:

(a) The early 'primary' components of evoked potentials are elicited by a cortical volley which projects mainly or exclusively to the primary cortical areas of the side opposite to the side of the body where sensory receptors are activated. The early SEP components are an example of this situation (fig. 1A, B). For the auditory evoked potentials there is still some discussion about which is the earliest cortical event [cf. JEWETT et al., 1970; GALAMBOS and HECOX, 1977], but special procedures have recently been developed along that line [PERONNET and MICHEL, 1977]. These early components can thus be lateralized for anatomical reasons, independently of any cognitive task being carried out, and they are indeed not changed by task conditions [DESMEDT et al., 1965, 1977; DEBECKER and DESMEDT, 1966; DESMEDT and ROBERTSON, 1977].

(b) The 'middle range' ERP components (about 70–200 msec latency) which have recently been shown to be specifically influenced by certain task conditions. For the somatosensory modality it has been shown that certain such components were lateralized to a greater or less extent to the hemisphere contralateral to the sensory receptors activated (fig. 2–4).

(c) The later ERP components which can be enhanced by certain task conditions and of which the best known example is the P300 components. The present paper has documented the symmetrical distribution of P300 (fig. 1, 4B).

(d) The ERP components which can be elicited or enhanced by certain task contitions and which are lateralized, not in relation to the body side where sensory receptors are being activated, but according to the hemisphere specialization for example in relation to language or nonlanguage processing behavior. Such transient ERP components are of course the most pertinent in relation to ERP studies of hemispheric lateralization, but only few examples have been critically documented to-date. We think that the active touch positive electrogenesis recorded over the right parietal area (fig. 5) represents one of the most striking instances of such a focalized phenomenon.

Some comments about this proposed differentiation of ERP components are in order. The early components considered under (a) have not been modified by selective tasks which suggests that afferent transmission in the lemniscal pathways leading to the primary cortical areas was not modulated significantly by the paradigms considered [DESMEDT et al., 1965; 1977; DESMEDT and ROBERTSON, 1977]. This of course does not rule out the possibility that some control of afferent transmission at the thalamic level would occur along the lines suggested by SKINNER and YINGLING [1977], and such a mechanism might indeed very well influence later components of the ERP. This is a question which remains for future studies.

The 'intermediate components' quoted under (b) are quite interesting and they have only recently been studied in detail. Such components, of which the N150 is an example, have been recorded in bisensory paradigms in which they seem to be related to the ERPs of the attended channel (fig. 2) [cf. DESMEDT and DEBECKER, 1978]. We were surprised to find that for intramodality selective attention, in the slow four-finger paradigm, these components were left virtually unaffected [DESMEDT et al., 1977]. Indeed it appears that a task involving a speed overload may be required to elicit a N150 enhancement in the intramodality selective attention type of experiment, with simple (and nonnatural) stimuli.

When a selective attention task must be carried out at a faster rate, its cognitive load can be considerably increased and the performance can actually be disrupted. In forced paced psychological experiments involving a simple one-bit decision about target and nontarget acoustic clicks (20–30 dB difference in intensity) the rate of correct decision did not exceed a ceiling of 2.45/sec at high rate [DEBECKER and DESMEDT, 1970]. The ERPs to target clicks were studied for different rates of presentation: at slow rates such as 1/sec the ERPs presented little, if any, decision component as the task was apparently too easy to engage the subject's processing resources; however, when the rate of presentation was increased to 2–5/sec, a large negative component with a peak at about 140 msec was regularly elicited [DEBECKER and DESMEDT, 1971]. These data suggested that the forced paced conditions represent by themselves a cognitive overload, even when simple decisions are involved which required little processing at slow rates. Thus 'to some extent the intrinsic difficulty of a discrimination and the speed at which it has to be made can be traded' [DEBECKER and DESMEDT, 1971, p. 120].

This approach which constrains the subject's processing resources and elicits additional cognitive ERP components through a speed overload has been emphasized by HILLYARD et al. [1973] who used four different acoustic stimuli presented over the two ears and who varied the rate of delivery; the experimental design was such that the subject was forced to attend one channel (ear) at a time [SCHWENT and HILLYARD, 1975; SCHWENT et al., 1976]. Selective listening to stimuli to one ear elicited an increase of the corresponding N100 component and a decrease of the N100 in the ERPs evoked by from the opposite ear. Only the target stimuli to the attended channel elicited a P300. The effect of task difficulty and delivery rate of the acoustic stimuli on the appearance of N100 was disclosed in experiments carried out at various rates, thus confirming the relation of N100 increase to speed overload under these conditions [SCHWENT et al., 1976; HILLYARD and PICTON, 1978]. Unfortunately these data were only recorded from the midline and the lateral distribution of N100 to acoustic stimuli has not been investigated.

Our recent work on forced paced selective attention tasks in the somatosensory modality has disclosed a clear lateralization of these intermediate ERP components, both for N150 and P100 [DESMEDT and ROBERTSON, 1977]. Moreover we also found that N150 enhancement could be associated both to the target finger stimuli and to the nontarget stimuli delivered to the finger adjacent to the one attended in the same hand. This contrasted with the P300 which only increased for the target stimuli [DESMEDT and ROBERTSON, 1977]. We concur with the views of HILLYARD et al. [1973] and SCHWENT et al. [1976] to suggest that the increase of intermediate ERP components in forced paced tasks reflects the activity of a special processor function dealing with afferent signals arriving over the attended channel (in our experiments, arriving from the same hand). Introspection indeed suggests that, while stimuli to the opposite nonattended hand can relatively easily be attenuated or even wiped out of consciousness, the task of telling apart target from nontarget signals in the same attended hand can be rather difficult at fast rates. It is thus not surprising that special processing strategies would be switched in, in order to permit differentiation between these signals in order to achieve proper identification of the targets. The situation corresponds to the 'stimulus set' condition described for psychological experiments of similar design by BROADBENT and GREGORY [1964] and TREISMAN [1969] [cf. also HILLYARD and PICTON, 1978].

The special 'stimulus set' processing thus considered apparently involved both hemispheres, but with a characteristic lateralization to the side contralateral to the body side receiving the target stimuli. Such lateral prevalence is perhaps not surprising since the afferent target signals from the hand arrive at the opposite primary cortical areas which are heavily connected anatomically with the posterior parietal associative areas on the same side [cf. JONES and POWELL, 1973]. That the intermediate ERP components should in most cases not be entirely contralateral to the side stimulated is not surprising either, in view of the important callosal connections between the parietal associative areas.

The later components and particularly the P300 have an entirely different significance since P300 is associated with target stimuli only [TUETING *et al.*, 1971; HILLYARD *et al.*, 1973; HILLYARD and PICTON, 1978]. We were at first surprised that P300 apparently maintained a symmetrical distribution even in the fast four-finger paradigm where the intermediate ERP components were lateralized (fig. 4). FRIEDMAN *et al.* [this volume] have used P300 in experiments on ERP language correlations; they found that P300 was enhanced to the last word of a sentence which they interpret as a correlate of syntactic closure. We think that P300 does not reflect the actual processing functions for language or nonlanguage material in the ERP experiments dealing with higher cognitive functions. Rather we would favor the view that P300 indexes the completion of a perceptual decision and the clearance of the information channel for other activities. Such an interpretation would be in line with the symmetrical distribution of P300, even in tasks where asymmetrical involvement of the two hemispheres is likely to occur [DESMEDT and ROBERTSON, 1977].

Except for signalling such closure function, P300 might thus turn out to be relatively uninteresting and it would not provide the much expected tool for probing hemispheric specialization. Along these views we can remain cold-blooded when hearing catastrophically negative statements about ERP and P300 research potentialities in this field.

Finally the experiments quoted under (d) above are of course the ones to be brought into focus by current work on hemispheric specialization. Steady attempts to engage the special hemispheric processors for language or for nonlanguage functions are only beginning to provide adequately controlled and substantial data. The problem is of course to design the appropriate experimental paradigms which will be compatible with transient ERP averaging while being sufficiently sophisticated and meaningful to engage the special processor considered. The present evidence of a specific prolonged positive electrogenesis recorded from the right hemisphere during active touch spatial orientation tasks seems to provide a highly promising line (fig. 5) [DESMEDT, 1977d].

Prof. JOHN E. DESMEDT, Brain Research Unit, University of Brussels, Boulevard de Waterloo 115, *B–1000 Brussels* (Belgium). Tel. (02) 538 08 44.

Language and Hemispheric Specialization in Man: Cerebral ERPs.
Prog. clin. Neurophysiol., vol. 3, Ed. J. E. DESMEDT, pp. 188–204 (Karger, Basel 1977)

The Ontogeny of Cerebral Asymmetry in Man: Auditory Evoked Potentials to Linguistic and Non-Linguistic Stimuli in Infants and Children

DENNIS L. MOLFESE

Department of Psychology, Southern Illinois University at Carbondale, Carbondale, Ill.

Since the time of WERNICKE [1874], there has been a growing belief that the 2 hemispheres of the human brain have quite different cognitive functions. That language functions are controlled by mechanisms primarily within the left cerebral hemisphere has been documented by work with brain-damaged individuals with techniques such as the study of patients with missile wounds [NEWCOMBE, 1969], cortico-vascular accidents [GESCHWIND, 1965], split-brain surgery [GAZZANIGA, 1970] or hemispherectomies [BASSER, 1962] and patients studied by cortical brain stimulation [PENFIELD and ROBERTS, 1959]. Only within the last 2 decades have techniques been developed which permit the study of brain-language functions in neurologically intact individuals. These methods have included the use of dichotic listening procedures [KIMURA, 1961, 1964, 1966] and, more recently, the use of evoked potentials (EP). Such procedures are not exclusively limited to adults possessing a mature language system nor are they restricted to clinical patients with brain abnormalities. In addition, these techniques enable investigators to assess language abilities in very young children who are unable to make appropriate physical responses (e.g. verbalizations, gestures, etc...). To some extent, the use of auditory evoked potentials (AEP) has supported the findings of researchers who had previously worked with brain-trauma cases. Hemispheric differences in the AEP to language materials have been reported in adults [COHN, 1971; MCADAM and WHITAKER, 1971; WOOD et al., 1971; MORRELL and SALAMY, 1971; MOLFESE, 1972; MATSUMIYA et al., 1972; TEYLER et al., 1973]. However, in several of these studies the findings have sometimes appeared to be inconstant and, when present, the hemispheric differences in voltage of EP were generally very small (less than 2 μV).

Experimental research on the ontogeny of the lateralization of language functions is as yet lacking. On the basis of clinical reports, LENNEBERG [1967, 1973] postulated that lateralization developed between 20 and 36 months of age. However, infants (like adults) discriminate between speech sounds on the basis of the phoneme category rather than only on the basis of simple acoustic changes within the sound [EIMAS et al., 1971; TREUB, 1973]. These recent findings suggest that some areas of the brain may have already specialized to process various types of language materials very early in life. In addition, anatomical studies of premature and full-term infants have suggested that some differences between hemispheres are present in early life. WADA et al. [1975] reported the presence of morphological asymmetries in the frontal operculum and the temporal planum of fetal infants as early as 29 weeks gestational age. WITELSON and PALLIE [1973] have also reported such asymmetries in both premature and full-term infants. These hemispheric differences appear to be at least as great as those reported for adult brains [GESCHWIND and LEVITSKY, 1968; WITELSON and PALLIE, 1973]. Given these behavioral and anatomical findings, it appears that some components of the language system may be operational in infants during the first months of life.

This chapter attempts to assess the ontogeny of hemispheric differences for language materials from early in infancy and to isolate some of the acoustic factors responsible for asymmetrical responding. The first study investigated developmental changes in hemispheric asymmetry.

I. Developmental Changes in EP Asymmetry

Methods

31 subjects were divided into 3 age groups. The adult group included 5 male and 5 female graduate students between 23 and 29 years (mean = 25.9 years). A group of 11 children included 5 males and 6 females between 4 and 11 years (mean age = 6 years). The infant group consisted of 10 infants, 4 males and 6 females, who ranged from 1 week to 10 months of age (mean age = 5.8 months). The infants and children were selected from birth announcements in a local paper and from cooperative nursery school lists. Parents were contacted by letter and by phone, and asked to permit their children to participate in the study.

None of the subjects had prior experience with AEP recording procedures. Handedness of the subjects, though not controlled, was recorded. Nine of 10 adults reported that they used their right hand to a greater extent than their left hand to open doors, throw a ball, hold a fork and write. Ten of 11 children responded that they were right-handed and their parents concurred. Since handedness was not testable in infants, parental hand preferences were recorded. 17 of 20 parents were classified as right-handed. No infant had 2 left-handed parents.

All subjects were seated in a black vinyl recliner chair positioned within a copper-shielded cage ($2.4 \times 3.6 \times 2.4$ m) which was electrically grounded and located at one end of a carpeted, sound-dampened room. The room temperature during the experiment was maintained at 18.5 °C. The subject's field of vision immediately in front of the chair was homogeneous, consisting of one side of the copper cage. The speaker through which the stimuli were presented was positioned on the outside of the shielded cage, approximately one m above the subject's head. The infants were held in a reclining position in their mother's laps and their heads were immobilized facing the speaker. The left and right temples were rubbed with fine grained sandpaper, washed with alcohol and received Grass silver electrodes (E5S) which were attached by means of Grass electrode paste. A rubber head band was used to hold the electrodes in place. The electrodes were placed at the Ten-Twenty International Electrode System temporal locations T_3 and T_4. Linked ear reference electrodes were attached with paste and ear clips. Skin resistance between electrodes was maintained under 7 kΩ and equal within 1kΩ for both sides. Impedances were checked periodically during the session. The left and right AEPs were amplified with Tektronix type 3A9 differential amplifiers with band-pass filters flat between 1 and 100 Hz.

All monitoring and recording equipment was located in a room immediately next to the experimental room. The AEPs were recorded on 0.5-inch magnetic tape by a 7-channel Ampex FM Recorder (Model PR 500). Recorded simultaneously was a 1.2 V, 400-Hz sine wave triggered by a Grason-Stadler voice-operated relay (Model 37300A-1) with a 10-msec operation time which, in turn, was keyed by the onset of the auditory stimulus. The 400-Hz signals served to program the subsequent averaging. Data from the tape were fed into a PDP-8 computer, digitized with 1-msec bin width and averaged with an analysis time of 750 msec for adults and 1,500 msec for infants and children.

Once the electrode preparations were completed, the adults and children were instructed to remain as motionless as possible and to attend to the sounds played through the overhead speaker. The mothers of the infants were asked to help their children remain as quiet and motionless as possible during the tests. If an infant appeared to fall asleep during the testing session, the session was temporarily suspended until the infant was awakened. Six sounds served as stimuli: 2 non-meaningful speech syllables /ba/ and /da/, 2 mechanically produced sounds which consisted of a C-major piano chord produced on a Steinway upright piano and a speech noise burst which was composed of frequencies 250 Hz to 4 kHz generated by a Grason-Stadler noise generator (Model 901B), and 2 words /bɔi/ and /dɔg/. The verbal stimuli were spoken by an adult male. The non-language sounds were selected to serve as control stimuli.

All stimuli were initially recorded at equal peak intensity levels on one channel of a Sony stereo tape recorder (Model TC560). The sounds were then edited so that all were 500 msec in duration. The editing did not interfere with the perceptability of the sounds. The 6 stimulus sounds were then spliced on a reel of tape and recorded again on 1 channel of a Panasonic stereo tape recorder (Model RS 7558). Five additional orderings of the stimuli were constructed by resplicing the stimuli in various orders and then recording them on a second tape recorder. In this manner a 6×6 Latin square, composed of 6 different orderings of the 6 stimuli was constructed. The time interval between stimuli was varied randomly between 2.5 and 4.5 sec to further reduce the possibility that the subject would habituate to the stimuli. A keying signal was recorded on the second channel consisting of a series of 200, 400, 600, 800, 1,000, and 1,200 Hz sine waves generated by a Precision

Paco Signal Generator (Model E-310). One of the 6 frequencies was paired with each of the stimuli on channel one. For example, the 200-Hz sine wave was always paired with the speech syllable /ba/. The entire sequence was then recorded on the Sony stereo tape recorder 20 times so that one tape had 20 continuous repetitions of the 6×6 Latin square. The sounds on channel one served as the stimuli for the experiment and were presented to each subject at 80 dB SPL. The keying signals were recorded on the technical channel of the Ampex FM recorder during the experiment to allow subsequent off-line pooling of similar AEPs in separate averages.

Results

The most consistently observed features of the AEPs were 2 large and temporally reliable components. The largest shifts in potential occurred between 454 and 528 msec in infants, 143 and 214 msec for the children and 90 and 160 msec in the adults. The latencies of these components appeared to decrease with maturation. This decrease in the latencies of EP components with age is consistent with findings from other studies [PRICE and GOLDSTEIN, 1966; ORNITZ et al., 1967; SUZUKI and TAGUCHI, 1968; DESMEDT, 1971; OHLRICH and BARNET, 1972; DESMEDT et al., 1976]. The largest negative (N) and positive (P) shifts in the AEP were selected for analysis. These components were measured in adult in other studies of laterality differences [MATSU-MIYA et al., 1972; MORRELL and SALAMY, 1971]. The latencies of these peaks were approximately equal for all the stimuli used, within each age group. Amplitude (NP) was measured from the peak of the major negative wave (N) to the peak of the major positive wave (P). An index of laterality of AEPs was obtained by dividing the NP amplitude for the left hemisphere by the summed NP amplitudes for both hemispheres. An R-value larger or smaller than 0.5 indicates a larger amplitude of the NP wave in the left or right hemisphere, respectively.

An analysis of variance showed that the 2-way interaction of groups (infants, children, and adults) and sound type (mechanical sounds, speech syllabes and words) was significant ($F = 9.45$; d.f. $= 4$; $p < 0.01$). Also significant were the main factors of groups ($F = 23.54$; d.f. $= 2$; $p < 0.01$) and sound type ($F = 141.06$; d.f. $= 2$; $p < 0.01$). Orthogonal contrasts of the means involved in the significant interaction were calculated [WINER, 1962] and significance levels were established using the Scheffe method. These contrasts indicated that the mechanical stimuli elicited proportionately greater AEP asymmetry in infants than in children and adults ($p < 0.01$). The children and the adults responded to these stimuli with the same degree of right hemisphere asymmetry. The speech syllables elicited approximately equal degrees of asymmetry in the infants and children although both groups

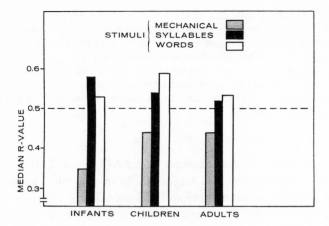

Fig. 1. Mean R-value of the infants, children, and adults in response to the speech syllables, words and mechanically produced stimuli. An R-value greater than 0.50 indicates a greater amplitude AEP in the left hemisphere while an R-value below 0.50 indicates a greater amplitude AEP in the right hemisphere.

responded with greater asymmetry than did the adults (p<0.01). However, the words elicited greater asymmetrical AEP activity in the children than in the infants and adults (p<0.01).

The infants', childrens', and adults' median R-value responses to the 3 stimulus classes are shown in figure 1. Note that the amplitude differences in the infants for both the syllables and the mechanical stimuli appears to be much greater than that seen in AEPs of the children and adults. It is interesting to note that the largest hemisphere differences in the adults occur in AEPs to the mechanical stimuli. The AEP to the verbal stimuli was greater in the left hemisphere in 9 of the 10 infants while the EPs to the piano chord and noise were greater in the right hemisphere of all 10 infants. Ten of 11 children responded with greater left hemisphere AEP to the verbal stimuli, 9 children showed greater right hemisphere AEP to the noise and all children responded to the piano chord with greater right hemisphere AEP. Eight of 10 adults showed greater left hemisphere AEP to the verbal stimuli. Nine of the adults responded to the piano chord with greater right hemisphere AEP and all 10 adults responded with greater hemisphere AEP to the noise.

The AEPs of 2 infants to the piano chord are presented in figure 2. The maximum NP differences in both infants are greater in the right hemisphere for both subjects. The AEPs from the same 2 infants to a speech syllable are presented in figure 3. In this case, the largest NP differences occur in the left

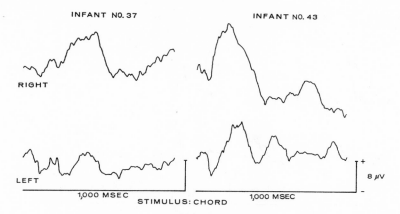

Fig. 2. Averaged EPs to the musical chord stimulus for 2 infants. The top and bottom traces for each infant are from the right (T4) and left (T3) hemispheres, respectively. Time scale, 1 sec; vertical calibration, 8 μV.

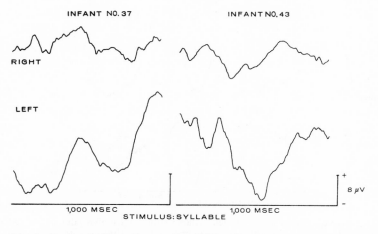

Fig. 3. Averaged EPs to the speech syllable stimulus for two infants. The top and bottom traces for each infant are from the right (T4) and left (T3) hemispheres, respectively. Time scale, 1 sec; vertical calibration, 8 μV.

hemisphere of both infants. The magnitude of the hemisphere differences for the infants in response to the various stimulus conditions averaged approximately 8μV across all infants. Adult AEPs to the musical chord and the speech syllable are presented in figures 4 and 5, respectively. In figure 4, the left-right hemisphere differences for subject 8 is approximately 3 μV, while

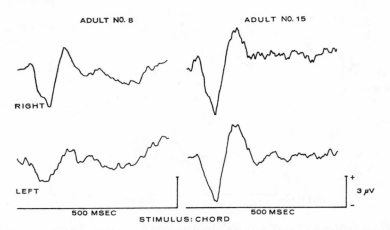

Fig. 4. Averaged EPs to the musical chord stimulus for 2 adults. The top and bottom traces for each adult are from the right (T4) and left (T3) hemisphere, respectively. Time scale, 500 msec; vertical calibration, 3 μV.

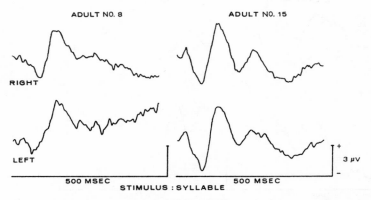

Fig. 5. Averaged EPs to the speech syllable stimulus for two adults. The top and bottom traces for each adult are from the right (T4) and left (T3) hemispheres, respectively. Time scale, 500 msec; vertical calibration, 3 μV.

it is only 1 μV for subject 15. Both subjects responded to the piano chord with greater amplitude right hemisphere AEPs. However, the same 2 subjects responded to the syllable stimulus with greater left hemisphere AEPs. The left hemisphere AEP amplitude was approximately 0.75 μV greater than the right AEP for subject 8 and only 0.25 μV greater in subject 15. As indicated, 27 subjects responded to the verbal stimuli with larger left hemisphere AEPs in comparison with their right hemisphere AEPs. None of the children or adults who responded with greater right hemisphere AEPs to the verbal stimuli were left-handed while the one infant who responded in this manner did have a mother who was left-handed.

Discussion

The results indicate that even prior to 1 year after birth, the human brain has specialized to such an extent that AEPs to language stimuli appear to be lateralized to a greater extent to the left hemisphere while responses to non-language stimuli are lateralized to a comparable extent, if not more so, to the right hemisphere. When the trends in lateralization for the various stimuli are compared across the 3 age groups, 27 of the 31 subjects responded to the verbal stimuli with greater left hemisphere AEPs, while 30 subjects responded to the piano chord and 28 subjects responded to the noise stimulus with greater right hemisphere AEPs. The principal hemisphere for responding to particular stimuli remained constant across groups with greater amplitude left hemisphere AEPs occurring to the verbal materials and greater amplitude right hemisphere AEPs occurring to the mechanical stimuli. The existence of the larger left hemisphere AEPs to words and syllables in infants less than 1 year old and their similarities to the AEPs in children and adults suggest that some of the cortical processes involved in the perception of speech may be present in pre-verbal infants. Of significance also is the fact that the lateralization of AEPs appears to be particularly strong for non-language sounds to the right hemisphere.

The differences in the infant AEPs to the various stimuli do not appear to be the result of simple frequency differences among these sounds. Both the noise and the verbal stimuli contained complex frequencies of sound between 250 Hz and 4 kHz. If an infant was responding only to the frequencies of the sounds and not to specific organization, he would have responded to both the noise and the verbal stimuli with comparable levels of AEP activity in the same hemisphere. However, this was not the case. Rather, some organization of speech frequencies must be the determining factor for the lateralization of verbal materials. Whatever this organization is, it is different from the organization of the stimuli to which the right hemisphere responds.

The magnitude of the hemisphere differences in adults are comparable to those reported by others [MORRELL and SALAMY, 1971; TEYLER et al., 1973; see also this volume] averaging of 1.25 μV for the verbal stimuli and of 2.8 μV for the mechanical stimuli. However, both the AEP amplitudes and the differences between the hemispheres remained relatively small. For the infants, the degree of EP asymmetry and the magnitude of the differences between hemispheres were greater than for adults: average hemisphere differences were 8 μV for verbal stimuli and 16 μV for mechanical stimuli. For all subjects, the hemisphere differences were greater for the mechanical stimuli than for the verbal stimuli.

That developmental changes in hemispheric responding do occur is shown by the fact that latencies of AEPs and degree of lateralization to both verbal and mechanical stimuli decrease with increasing age. One possible explanation as to why lateralization of AEPs declines with age may be an increasing interaction between the 2 hemispheres, as a result of the maturation of the corpus callosum and various commissures. Myelination of the commissures is not complete at birth and continues through to puberty [CONEL, 1941, 1967; SELNES, 1974]. HEWITT [1962] noted that the corpus callosum is proportionately thinner in neonates than in adults and that parts of the callosum, such as the rostrum and splenium, continue to develop later. With increasing myelination and development there must be an increase in the interaction between the 2 hemispheres. Other mechanisms may be involved as well in the reduction with maturation of the AEP disparity found between the 2 hemispheres.

The results of the present study have interesting implications for LENNE-BERG's [1967] suggestion that the ability of a patient to recover language following a severe cortical vascular accident (CVA) to his dominant hemisphere may be a function of the degree to which language has been lateralized to that hemisphere. He further suggested that a CVA between 20 and 36 months of age might allow recovery of 'perfect speech' abilities since lateralization has not yet occurred, while a CVA in the language-dominant hemisphere after lateralization has occurred is not followed by recovery of the lost language skills. Since the present study indicates that language lateralization is already present in infancy, it becomes difficult to accept that the state of lateralization acquired would determine the recovery of language functions after CVA. Rather, it appears that some other factor, such as brain plasticity which is thought to decrease with age, might be the decisive factor in the recovery of language abilities.

Thus, developmental changes in hemispheric responding occur between infancy and adulthood. At all ages, however, the hemispheres appear to be differentially sensitive to some acoustical aspects of stimuli. The next step in research is to determine the characteristics of stimuli which trigger asymmetrical electrical responses.

II. *Acoustic and Linguistic Factors Relevant to EP Asymmetry*

In adults, the asymmetry of EPs to auditory stimuli may be due to 2 factors: linguistic meaningfulness and/or acoustic characteristics. The effects of the first factor were investigated by PAPCUN *et al.* [1974], who used a dichotic listening procedure to present morse code signals either to morse code instructors or to morse code naive college students. He found no ear differences in the college population, but the instructors reported the signals with a significantly higher degree of accuracy when the sounds were presented to the right ear. Since materials presented to the right ear are thought to be processed primarily in the left hemisphere [ROSENZWEIG, 1951; KIMURA, 1961], it would appear that the instructors were better able to process the coded signals in their left hemispheres. Since these sounds were radically different from speech stimuli, PAPCUN *et al.* argued that the left hemisphere may process not auditory signals *per se,* but materials which have some degree of linguistic meaningfulness.

CUTTING [1974] presented a strong case which suggests that the left hemisphere may also attend to specific acoustic elements. Using a dichotic listening procedure with adults, CUTTING presented a series of vowels and consonant-vocel syllables which were synthesized either with a narrow bandwidth (non-speech) or with a normal bandwidth (speech). The non-speech stimuli were constructed from sine waves which corresponded exactly to the central frequencies of the formants in the speech stimuli. The bandwidth of these non-speech stimuli was essentially zero. In contrast, the bandwidth of the formants for the speech stimuli were 60, 90, and 120 Hz for F_1, F_2, and F_3, respectively. The stimuli also varied according to transition. The vowel sounds were steady state and did not contain a transition; the con-sonant-vowel syllables did contain a rapid frequency change. CUTTING [1974] found that the speech materials produced a right ear advantage while the non-speech stimuli did not. The presence of transitional elements also appeared to result in a right ear advantage, although the magnitude of this difference was not as large. In fact, the transition effect appeared to be largely due to the large right ear advantage of the speech consonant-vowel (16%). The non-speech, consonant-vowel stimuli only resulted in a 1% advantage for the right ear.

For adults, then the characteristics of the sounds themselves play a role in determining whether a sound is to be processed in the left or the right hemisphere. However, as indicated in the work by PAPCUN [1974], semantic or cognitive factors also appear to be at work. These findings present an interesting problem with respect to asymmetrical responding in infancy. Infants at birth are generally thought to have only a reflex based cognitive system [PIAGET, 1970] and little or no linguistic skills [MACNAMARA, 1972; TREVARTHEN, 1974]. Therefore, if neonates respond asymmetrically to speech stimuli, these asymmetrical responses must be based on acoustic components of the stimuli rather than the linguistic factors. Perhaps at a later point, when the infant acquires language skills, the left hemisphere would be responsive to either acoustic or linguistic elements. Initially, asymmetrical responding may be triggered only by acoustic elements. The present study attempts to isolate the acoustic factors which elicit asymmetrical responding in neonates.

Method

14 neonates (7 males and 7 females) were tested within 24 h of birth. The mothers received either no or minimal anesthesia during delivery.

Five artificially produced sounds were used as stimuli: 2 speech syllables, two non-speech syllables, and a 500-Hz tone. The speech and non-speech stimuli were constructed by CUTTING at the Haskins Laboratories. The speech syllables were constructed on a parallel resonance synthesizer and included a consonant-vowel (CV) syllable /gae/ and a steady state vowel (V) syllable /ae/. The speech stimuli were composed of 3 formants and all had the same fundamental frequency of 110 Hz. The bandwidth of the speech stimuli were 60, 90, and 120 Hz for F_1, F_2, and F_3, respectively. The bandwidth for the nonspeech stimuli was 1 Hz for each formant. Non-speech stimuli were composed of 3 tones which corresponded exactly to the central frequency of the formants in the speech stimuli: 660, 1,720, and 2,410 Hz. The formants of one non-speech stimulus (/ae/) consisted of steady state sine wave segments, while the second non-speech stimulus (/gae/) was composed of an initial 50-msec frequency modulation followed by the steady state segment. The modulation corresponded to the transition in the CV speech stimulus. The third non-speech stimulus, the tone, consisted of one 500-Hz sine wave. All stimuli were identical in peak intensity and were 300 msec in duration.

40 random orderings of the 5 stimuli were recorded on one channel of a stereotape recorder. The time interval between speech sounds varied between 4 and 8 sec. A 50-Hz, 0.5-V square wave was recorded on the second channel of the tape recorder 50 msec before the stimulus on channel 1. This pulse signaled a PDP-12 computer to begin averaging. Each infant was tested individually. Infants were initially swaddled in a blanket to prevent their hands from dislodging the recording electrodes. Each infant was then placed into an infant seat which was reclined at a 20° angle. Recording electrodes were placed on the left (T_3) and right (T_4) temporal scalp locations, and 2 linked electrodes were attached to the ear lobes to serve as references for the temporal leads. A fifth electrode was placed on the forehead over the right eye to monitor eye movements. Resistance of all electrodes was checked throughout the study and maintained under 3 kΩ. All electrodes were plugged into an Analogue Devices isolation unit to isolate the infant in the event of equipment failure. The output of the isolation unit was connected to 3 Tektronix AM 502 differential amplifiers. The amplified signals were then recorded on a Vetter C-4 cassette FM recorder. Recorded data were later averaged and analyzed on a PDP-12 computer with a modified

version of Reeder's averager (DEC Corporation). During the recording, a female research assistant monitored the infants' movements. In the event that the infant cried or pulled on an electrode, recording was temporarily halted until the behavioral or technical problems were corrected. Testing sessions varied in length between 40 and 75 min.

Results

The neonates' AEPs were characterized by a large positive-going waveform (P_{250}) which reached a peak approximately 250 msec (mean value) following stimulus onset. This, in turn, was followed by a negative-going waveform (N_{540}) which peaked 540 msec (mean value) after stimulus onset. These 2 peaks, which marked the largest shift in potential, were selected for analysis and measured by a PDP-12 computer programmed to search for a large positive-to-negative shift between 200 and 600 msec following stimulus onset. These amplitude measures were then converted into R-values.

An analysis of variance showed a significant effect for speech versus non-speech stimuli ($F = 17.73$; d.f. $= 1, 13$; $p < 0.01$). The infants responded to the speech stimuli (broad bandwidth) with a greater amplitude left hemisphere response, while the non-speech stimuli elicited a greater amplitude right hemisphere AEP. The AEPs from 2 infants in response to a speech stimulus are given in figure 6. Note that the left hemisphere response of both

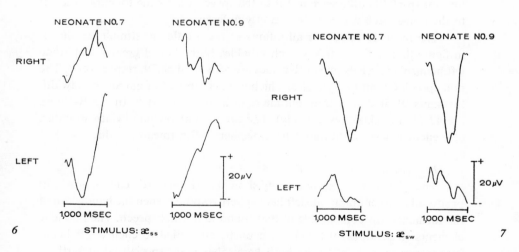

Fig. 6. AEPs to a speech stimulus for 2 infants. Time scale, 1 sec; vertical calibration, 20 μV.

Fig. 7. AEPs to a non-speech stimulus for 2 infants. Time scale, 1 sec; vertical calibration, 20 μV.

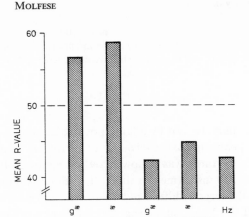

Fig. 8. Means R-values for 14 neonates to 2 speech and 3 non-speech stimuli.

infants is larger than that for the right hemisphere. Figure 7 illustrates the AEPs for the same 2 infants to the non-speech stimuli which were identical to the stimuli in figure 6 except for bandwidth. Here the amplitude of the right hemisphere is larger than that for the left hemisphere. Comparisons of the tone stimulus with the speech and non-speech stimuli showed a significant ($p < 0.01$) difference in EP to the speech versus the tone, but the EP to the non-speech was not significantly different.

The mean R-values for all infants to the 5 different stimuli are shown in figure 8. The CV and V speech syllables both elicited greater amplitude left hemisphere AEPs (mean R-values are 56.64 and 58.79, respectively). The non-speech CV and V syllables, which were composed of narrow bandwidth formants, elicited greater amplitude right hemisphere AEPs (mean R-values are 42.43 and 45.00, respectively). The tone, a narrow bandwidth stimulus, also elicited larger right than left hemisphere AEPs (mean R-value is 42.86).

Discussion

The results of this second study show that the acoustic cue (bandwidth) produced a larger amplitude left hemisphere response when the bandwidth of the formants was comparable to that found in normal speech. The phonetic elements (transitions) without the acoustic cue failed to elicit a left hemisphere response. Apparently, both hemispheres are specialized at birth to differentiate between different sounds. The left hemisphere appears to be particularly sensitive to bandwidth, but does not selectively respond to that cue alone. The noise stimulus in the first study was a broad bandwidth stimulus (250–4 kHz), yet it evoked a greater amplitude right hemisphere EP. Thus, some organization of the bandwidth appears to be critical. The right hemisphere, on the other hand, appears to respond to a variety of acoustic cues not found in speech – narrow bandwidth, extremely broad bandwidths, musical chords and pure tones. However, more research is needed to determine the specific characteristics of sounds which elicit greater amplitude right hemisphere EPs.

The finding that vowel sounds as well as consonants triggered greater left hemisphere AEPs is in disagreement with research using dichotic listening procedures to assess differential hemispheric responses [SHANKWEILER and STUDDERT-KENNEDY, 1967; STUDDERT-KENNEDY and SHANKWEILER, 1970]. These studies have found that the left hemisphere is better in identifying consonant sounds, but vowel sounds produce either a slight left hemisphere advantage or no differences between hemispheres.

The results are also at variance with CUTTING's [1974] findings concerning the effects of transitional elements. CUTTING reported that transitional elements produced a right ear (left hemisphere) advantage, even when the sounds contained formants which were only 1 Hz in width. When the data of the present study are grouped in terms of EPs to transitional (R-value = 49.54) versus non-transitional (R-value = 51.89) stimuli, there are no differences between hemispheres. If the data are grouped according to EPs to speech versus non-speech stimuli, there is greater left hemisphere responding to the speech stimuli (R-value = 57.71) and greater right hemisphere responding to the non-speech (R-value = 43.71). As indicated earlier, CUTTING's transitional effect may only be a spurious one. Non-speech transitional elements produced only a 1% advantage for the left hemisphere – a difference which is not likely to be significant if viewed in isolation.

Both discrepancies – those concerning vowel effects and those dealing with transitional effects – may be due to the differences in the techniques used to assess the hemispheric EPs. AEP procedures may be assessing quite different cognitive functions or physiological correlates of these functions than dichotic listening procedures. Likewise, different dichotic listening tasks, which make different demands on the subjects, may be assessing quite different functions [PISONI, 1971]. This, then, could raise serious questions concerning the validity of comparing findings concerning hemispheric functions when different techniques for assessing such functions are employed.

The differences between CUTTING's findings and our own may also be due to developmental factors. CUTTING used adult subjects in his experiments while the AEPs reported in the present study were recorded from neonates.

Developmental changes occurring between infancy and adulthood, as a result of neurophysiological and cognitive factors, might be reflected in changes in hemispheric responses. The notion that cerebral asymmetry develops into adulthood is a relatively old one [LENNEBERG, 1967, 1973] and, although our data suggest that the 2 cerebral hemispheres of man may function somewhat differently from birth, there is no reason to believe that further changes in cerebral asymmetry do not occur. The data from the first study indicating a difference in the degree of asymmetry between infants and adults seems to support this.

III. Evidence from Principal Component Analysis

A partial replication of experiment 2 [MOLFESE et al., 1976] involving the use of multivariate analysis procedures developed by CHAPMAN et al. [this volume] also identified portions of the AEP in neonates that were sensitive to transition and bandwidth changes. However, when the entire waveform was included in the analyses rather than only the two points (P_{250}–N_{540}) used in experiment 2, a radically different view of hemisphere differences emerged. MOLFESE et al. [1976] recorded AEPs from T_3 and T_4 scalp locations of 16 neonates. The averaged AEPs were submitted to a principal components analysis [DIXON, 1972] which identified 4 factors as accounting for 96% of the total variance. These factors reflected sex and hemisphere differences and were also sensitive to formant bandwidth and transition structure. However, no interactions occurred between cortical components sensitive to formant bandwidth or transitions and hemispheres. Both hemispheres apparently responded to changes in these speech cues in a similar fashion. The hemisphere effects that did emerge apparently were of a more general nature and perhaps reflected basic differences that were not differentially effected by specific stimulus characteristics manipulated in that study. Although both experiment 2 and the study of MOLFESE et al. found evidence of hemisphere differences in neonates, the nature of such differences reported are clearly different. The results of experiment 2 suggest that the hemisphere differences in neonates are stimulus related while the findings of MOLFESE et al. indicate general hemisphere differences unrelated to stimulus characteristics. When the responses from the individual subjects from these studies are examined, several interesting points emerge. First the responses of 8 of the 14 neonates in experiment 2 were inconsistent in the sense that 6 of the 14 neonates failed to differentially respond with greater right hemisphere amplitude to at least one of the two sinewave formant stimuli. Eight of 14 neonates responded to both broadband formant stimuli with reliably larger left hemisphere amplitude AEPs. Clearly there are marked inconsistencies in the individual responses of neonates when only two points were utilized in the analyses. Similar individual differences can be noted in experiment 1 [see MOLFESE et al., 1975]. Given such variability both within and between subjects, the use of amplitude measures based on a restricted set of points as a reliable measure of hemisphere differences must be subject to question. The use of the principal components analysis which analyzed all points in the waveform produced radically different results. The AEPs from all infants in the study of MOLFESE et al. were characterized by the general hemisphere component. Here, both group and individual data consistently reflected hemisphere differences.

The use of the principal components analysis imposes a different interpretation on the use of amplitude and latency measures as measures of hemispheric differences. The analysis involved in the study of MOLFESE et al. noted above isolated four factors characterized by different components that overlapped somewhat in time. These four factors add together with the centroid to produce the original waveform. In the case of the study of MOLFESE et al. [1976] where the factors isolated by the analysis were sensitive to sex differences, formant bandwidth, transitional cues and hemispheres, the amplitude differences between the left and right hemispheres for any restricted subset of points in the AEP may not be a 'pure' measure of hemisphere differences but rather are confounded by other stimulus and subject related responses. Such confoundings could account for some of the discrepancies in results between studies employing only a limited number of amplitude measures [FRIEDMAN, 1975].

Given the findings reported by MOLFESE et al. [1976], asymmetrical responding in the neonates may not be triggered by acoustic elements but rather may represent more general processing differences between the two hemispheres.

As noted earlier, changes in hemispheric differences may continue to change throughout development. Although initial asymmetries may be due to general differences between the two hemispheres, with maturation other factors such as acoustic, linguistic and cognitive processes may contribute to further differences between the hemispheres [PORTER and BERLIN, 1975]. Researchers studying changes in the perception of voice onset time (VOT) from birth to adulthood have recently identified such changes [MOLFESE, 1976; in press].

VOT, the relation between laryngeal pulsing and consonant release, is an important cue used in speech perception to differentiate between bilabial stop consonants such as /b/ and /p/ [LIBERMAN et al., 1967]. Adults can discriminate variations in VOT only to the extent that they can assign unique labels to the sounds. Listeners are unable to discriminate between VOT values of 0 and +20 msec (both are identified as /b/) or between VOT values of +40 and +60 msec (which are identified as /p/). However, they do discriminate between the two phoneme categories (/b/ from /p/). AEP procedures when coupled with the principle component analyses procedures of CHAPMAN et al. have identified electrophysiological correlates of such discriminations in the RH of adults [MOLFESE, in press], preschool aged children (MOLFESE and HESS, in press] and 2-month-old infants but not in neonates [MOLFESE and JONES-MOLFESE, in press]. In addition to changes in RH processing that appear to occur with age, the similarities across these studies would appear to underscore the reliability of the analysis procedures and phenomena briefly described here. All studies indicated the presence of general hemisphere differences which occurred independent of subject or stimulus variables. This effect is similar to the one reported by MOLFESE et al. [1976]. All studies also reported the presence of RH differences between males and females in the non-adult populations. The studies involving subject populations over 1 month of age identified a factor in both hemispheres that was sensitive to changes in VOT both within and between phoneme categories and a second factor with major latencies later in time than the first factor that responded to changes in VOT on the basis of phoneme category only in the RH. Such specific yet consistent findings across different subject populations are quite exciting and definitely warrant further investigation in order to more thoroughly assess the power of these new research tools.

Dr. DENNIS L. MOLFESE, Department of Psychology, Southern Illinois University at Carbondale, *Carbondale, IL 62901* (USA). Tel. (618) 536 2301.

Language and Hemispheric Specialization in Man: Cerebral ERPs.
Prog. clin. Neurophysiol., vol. 3, Ed. J.E. DESMEDT, pp. 205–211 (Karger, Basel 1977)

CNV and P300 Experimental Paradigms for the Study of Language

DAVID FRIEDMAN, RICHARD SIMSON, WALTER RITTER[1] and
ISABELLE RAPIN

The Saul R. Korey Department of Neurology and Rose F. Kennedy Center
for Research in Mental Retardation and Human Development, Albert Einstein
College of Medicine, Bronx, N.Y.

The question whether average evoked potential (EP) analysis in man
can provide data relevant to the cerebral localization of language function
has recently attracted much research interest. Most investigators in this area
have looked for EP amplitude asymmetry as a method of corroborating the
clinical and behavioral evidence that the left hemisphere is specialized for
linguistic decoding, while the right hemisphere is specialized for non-verbal
tasks [WANNER et al., this volume]. The asymmetry issue is still not settled:
COHN [1971], MATSUMIYA et al. [1972], MORRELL and SALAMY [1971], WOOD
[this volume], MOLFESE [this volume], MARSH and BROWN [this volume], and
ROEMER and TEYLER [this volume] have reported EP asymmetry to linguistic
and non-linguistic stimuli, while SHELBURNE [1972, 1973], GALAMBOS et al.
[1975], and ourselves did not find asymmetry. It also appears that some of
the early experiments in which asymmetries were found suffered from methodological flaws [FRIEDMAN et al., 1975a].

In addition to the reports of EP asymmetry, other investigators have
reported finding CNV asymmetries when subjects were cued as to whether
they would have to respond, for example, by pressing a key, to either a
verbal or non-verbal stimulus at S2 in the usual CNV paradigms [BUTLER
and GLASS, 1974a; MARSH and THOMPSON, 1972; LOW and FOX, this volume].
Since the CNV is a pre-stimulus event, the finding of asymmetry in this
situation suggests that the brain develops a set which is reflected in differential CNV lateralization depending upon whether the second stimulus is

---

[1] Also at Herbert H. Lehman College of the City University of New York, N.Y.

verbal or non-verbal. While the issue of hemispheric asymmetry and whether it is reflected in the EP is important, it obscures the equally critical area of the investigation of the EP correlates of semantic, syntactic, and phonologic information processing in language experiments.

We have designed a number of experiments, to be reviewed here, in which we have employed linguistic and non-linguistic stimuli and have not found consistent EP asymmetry. As a result, our research strategy has shifted to uncovering the EP correlates of semantic, syntactic, and phonologic information processing in experimental paradigms where linguistic stimuli are used. Since we had been investigating the variables which influence the P300 wave prior to our work with language, and because P300 has been shown by a number of investigators to be intimately related to cognitive information processing [DONCHIN and LINDSLEY, 1969; DESMEDT 1977b], our work with language focussed on the P300 wave, although we looked at the earlier potentials as well. The CNV has also been shown to be sensitive to the operation of cognitive variables [MCCALLUM and KNOTT, 1973; TECCE, 1977] and we included its study in some of our experimental paradigms. P300 has been shown to be generated in posterior temporo-parietal association cortex [VAUGHAN and RITTER, 1970], an area considered important for language function, since it is strategically located between the cortical association areas of the visual, auditory, and somesthetic systems [CHASE, 1972]. Our lateral electrodes were placed in this cortical association area midway between Pz and the mastoid process. All our experimental linguistic paradigms were designed to generate P300 waves and, at times, the CNV.

In the following visual experiments, the stimuli were 35-mm black on white transparencies projected via a Kodak Carousel projector (with an attached shutter permitting exposure times down to 5 msec) onto a piece of opaque plexiglass. The plexiglass was constantly illuminated so that there was only a slight change (20 ft-c) in luminous flux when the slide was projected. This insured that the information contained in the slide would contribute more to the EP than would the change in light flux. The visual angle with which our subjects viewed the stimuli varied depending upon the type of stimulus, but was always within his foveal vision. In the auditory experiments, the stimuli were real speech words and human sounds spoken through a voice key which provided a pulse upon word or sound onset. Stimuli were played back to the subject from tape over earphonees at a comfortable listeing level.

EPs were recorded in our early experiments by scalp electrodes at O_1 and O_2 since BUCHSBAUM and FEDIO [1969, 1970] had reported waveform differences of the VEP to linguistic and non-linguistic photic stimuli at these electrode sites, and/or by scalp electrodes placed midway between Pz and the mastoid process. Tektronix type 2A61 amplifiers set to pass a band between 0.06 and 50 Hz were used. The EEG and pulses corresponding to stimulus onset were recorded on 7-channel magnetic tape and read off-line into a LINC-8 computer where averages were obtained and written out on a CALCOMP X-Y plotter.

We began our experimentation by simply presenting language and and non-language stimuli where the subject had no other task than to attend. Our intention was to determine whether the perception of language stimuli *per se,* as reflected in the VEP, was different than the perception of non-language stimuli. In these experiments, we used a variety of visual stimuli, such as meaningful words, blank slides, scrambled words, lines differing in orientation, and random designs based on MILNER's [1971] nonsense figures test. In these experiments, with no task imposed on the subject, these diverse stimuli produced VEPs of similar morphology and latency, with no consistent asymmetry related to whether verbal or non-verbal stimuli had been presented.

In the previous experiments, we had no necessary assurance that our subjects were, in fact, processing the stimuli. We, therefore, required our subjects to respond to each stimulus in a simple reaction time design. However, aside from the production of small-amplitude P300s mainly in the parietal leads [RITTER et al., 1972], there were no differences with respect to the no-response experiments. In this last experiment, our subjects did not have to process semantic content in order to do the task. In order to make the response contingent on semantic content, we asked our subjects to perform a go, no-go reaction time task in a CNV experiment where they were cued as to whether the discrimination they would make at S2 would be verbal or non-verbal. This experiment followed the reports of MATSUMIYA et al. [1972] and MARSH and THOMPSON [1972] that 'meaningfulness' and 'set' affected both CNV and EP asymmetry.

The cueing stimuli were 1,000 and 1,030-Hz tones which either indicated that the subject would have to make an animal vs non-animal discrimination at S2, or a large-small amorphous form (modelled after MILNER's [1971] nonsense figures) discrimination at S2. The subject's task was the same in the control condition, where the tone was random with respect to S2, and thus conveyed no information to the subject. We also employed no-stimulus runs to determine if there were any pre-existing amplitude asymmetries in the resting EEG. The results, for one subject, are presented in figure 1. The data presented here are typical of the other subjects studied. In measuring amplitude of the EEG on the 2 sides of the head during the no-stimulus runs, we found no resting asymmetries. It is clear from figure 1 that tones which carried cue value evoked large-amplitude P300s, whereas those that did not have cue value did not evoke P300s. This is obviously related to the information-bearing nature of the tone in the cue condition [SUTTON et al., 1965, 1967; DESMEDT et al., 1965; see DONCHIN and LINDSLEY, 1969]. Since the tone informed the subject of the nature of the discrimination that he would have to make at S2, we expected lateral asymmetries to this stimulus. For subject G.N. (fig. 1), it appears that N100–P300 amplitude is larger on the right regardless of the type of cue. However, across subjects there were no consistent asymmetries found. CNVs following tones which carried cue value were consistently larger across subjects than CNVs following tones which did not have cue value; this can be clearly seen in the data of figure 1. Since the subject had to be prepared to respond to S2 in both conditions, and thus his level of attention would have had to remain the same, these CNV amplitude differences must be related to the information-bearing nature of the tone. Thus, we have demonstrated a correspondance between the P300 which precedes CNV and the CNV amplitude following this P300. The difference in CNV amplitude between the 2 conditions could possibly have been due to the cue

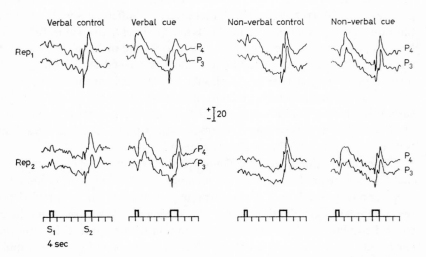

Fig. 1. EP and CNV data from a complete experimental protocol for subject GN, showing two replications (Rep₁ and Rep₂, n = 40 per replication) of the experiment.

allowing the subject to narrow his focus [TECCE and SCHEFF, 1969], whereas in the control condition the subject could not prepare to make a verbal or non-verbal discrimination, since the tone carried no cue value. Although the 2 conditions differed in the 'set' they created, as indeed reflected in the CNV amplitude differences, no lateral asymmetries across subjects were evidenced. We measured the N100–P300 component evoked by S2 and found no consistent asymmetries related to the verbal or non-verbal nature of the stimuli.

In another experiment where subjects had to process the stimuli for their semantic content in order to accomplish the task, we presented subjects with real speech words and human sounds [FRIEDMAN et al., 1975a]. We presented our subjects with 5 randomized words and 5 randomized sounds in separate blocks of 100 trials, where they had to respond with a finger lift to one signal stimulus and withhold their response to the four non-signal stimuli. A control condition (no task) in which the subject simply listened to the words and sounds was also run. We were thus able to collect EPs to the physically identical stimuli when they were signal, non-signal, and no-task stimuli. We found essentially no differences between words and sounds, but did find P300s to be present to all stimuli, even though non-signal stimuli do not usually evoke P300s [RITTER et al., 1972; HILLYARD et al., 1973]. The P300 varied in amplitude as a function of task demands being largest to signal, next largest to non-signal, and smallest to no-task stimuli. We speculated that the presence of P300 waves to all stimuli was due to the engagement of the P300 system whenever language stimuli were used and the subject had a task. Again, no consistent asymmetries to word or sound stimuli were detected.

Since most linguistic units occur in strings (either as sentences on the printed page or in connected discourse), we felt that putting words in a sentence where the subject had a task might produce VEP asymmetries. We, therefore, designed an experiment where visually flashed words comprised a sentence, in which a key word delivered task-related information either late (condition 1– last word delivered the information) or early (condition 2 – second word delivered the information) in the sentence [FRIEDMAN et al., 1975b]. There were no consistent VEP asymmetries in this study. We also found that all words in the sentence, whether redundant or not, evoked P300 components. However, regardless of when the task-related information was delivered, P300 to the *last* word in both conditions was significantly larger than the P300 to all the other words in the sentence, even though in condition 2, the second word delivered the information and the last word was redundant. We concluded that this enhancement of P300 to the last word of both conditions was a correlate of 'syntactic closure'. This 'closure' effect appears to be a unique EP correlate of syntactic linguistic processing. It is in this direction that EP research involving language could prove fruitful in uncovering EP correlates of linguistic processing.

P300 latency was found to be an important indicator of information processing in our language data. In the last experiment, P300 latency, and not P300 amplitude correlated with task-related information delivery. When a key word delivered information to the subject, P300 latency was longer to that word than to any other word in the sentence, or to the same position word in the other condition. In another experiment, we devised several categories of nouns (e.g. animals versus non-animals) and varied the size of the category (i.e. animals is a narrow category compared to non-animals which is a broad category). The subjects were asked to discriminate between words in the different categories using a choice reaction time paradigm. Both the reaction time and P300 latency lengthened as the size of the category increased, suggesting that memory search took longer the larger the category and that the EP could reflect this semantic processing. MARSH and BROWN [this volume] and CHAPMAN et al. [this volume] have also shown EP correlates of semantic processing. In an experiment designed to uncover EP correlates of phonologic processing, we asked our subjects to respond with a finger lift to a signal word among 4 words differing by one acoustic segment (pick, pack, pit, tick). P300 latency to the 4 words depended upon which word served as signal and was related to the time of occurrence of the critical segment in the word.

We have not found consistent asymmetry in tasks where the subject is cued to process verbal or non-verbal information (CNV paradigm), or in tasks where the stimuli delivered task-related information of a verbal or non-verbal nature. The literature dealing with EP asymmetry as a reflection of differential hemispheric specialization contains many inconsistencies in findings both within and between studies [FRIEDMAN et al., 1975a]. GALAMBOS et al. [1975] have recently undertaken studies similar to ours and to those of WOOD et al. [1971] and, after careful consideration of their data, have come to the same conclusions as we did. In addition, the literature concerning asymmetry of CNV as a reflection of verbal or non-verbal psychological set also contains contradictory findings: BRIX and BURIAN [1973], MARSH and THOMPSON [1973], and FRIEDMAN et al. [this paper] did not find asymmetry; while BUTLER and GLASS [1974a], MARSH and THOMPSON [1972], and DONCHIN et al. [1977b] did. The issue of hemispheric specialization and whether it is reflected in the EP is both theoretically and clinically important (e.g.

it could serve as a non-invasive, non-traumatic substitute for the Wada test). However, our findings and experimental paradigms suggest new ways in which P300 and CNV brain potentials can be utilized in the study of language.

Acknowledgments

The authors thank Dr. HERBERT VAUGHAN, jr. for the use of the LINC 8 computer and Ms. JULIE CHANG for the computer programs. The research reported was supported in part by Grants NS 2503, NS 3356, 2T1 N 55325 from the National Institute of Neurological Diseases and Stroke, and MH 06723 from the National Institute of Mental Health, United States Public Health Service.

Dr. D. FRIEDMAN, Department of Medical Genetics, Psychiatric Institute, 722 West 168th Street, *New York, NY 10032* (USA). Tel. (212) 568 4000, ext. 445.

Language and Hemispheric Specialization in Man: Cerebral ERPs.
Prog. clin. Neurophysiol., vol. 3, Ed. J. E. Desmedt, pp. 212–242 (Karger, Basel 1977)

Electroencephalographic Investigations of Hemispheric Specialization[1]

Emanuel Donchin, Gregory McCarthy and Marta Kutas

The Cognitive Psychophysiology Laboratory, University of Illinois, Champaign, Ill.

In this chapter, we review studies of the electrocortical manifestations of hemispheric specialization in humans – focusing on the relation between linguistic performance and hemispheric activities. In all of the reviewed studies, measures of the human electroencephalogram (EEG) or event-related potentials (ERPs) are used as dependent variables. This is an extensive revision of an earlier review which considered literature published before 1975 [Donchin et al., 1977]. In the present review, we have paid closer attention to the correlation between linguistic function and the electrical activity of the brain. Our previous review considered in some detail methodological problems which confront investigators in this area of research. This discussion is not repeated here and interested readers should consult the earlier paper.

Measures of various aspects of human EEG have long been used in clinical neurology [Cooper et al., 1974]. The scalp distribution of the EEG can be used in localizing epileptic foci [Gibbs et al., 1936], tumors [Walter, 1936], focal lesions [Case and Bucy, 1938], and other pathologies. Such success derives from the fact that these gross pathologies may be associated with localized electrical dipoles which generate interpretable distributions of activity at the scalp [Cooper et al., 1974]. The principal characteristic of such studies is that differences between activities recorded at different electrodes are used to infer intracranial events. The events of interest have

[1] Preparation of this report was supported, in part, by the Defense Advanced Research Projects Agency under Contract N–000–14–76–C–0002 with the Office of Naval Research. G. McCarthy is supported by a University of Illinois Fellowship. We thank George McKee for his comments on a preliminary draft.

well-defined physical properties, such that conclusions about them are usually based on sound physical principles. The papers reviewed here utilize the same logic – the scalp distribution of some EEG parameter is the source of inferences about the differential utilization of distinct cortical regions during the normal processing of information by the brain. The underlying assumption is that certain aspects of scalp-recorded activity are correlated with 'utilization'. Unfortunately, the physical properties of a 'utilized' population of cells are not very well understood and the effects such activities may have on scalp recording is even more in doubt.

The electrophysiological investigation of hemispheric specialization has blossomed with the increasing interest in the study of the complementary specialization of the cerebral hemispheres. Much evidence has accrued during the last two decades, demonstrating that the two hemispheres of the brain are not functionally equivalent. A grossly oversimplified summary of these data would consider the left hemisphere, in dextrals, as supporting verbal, analytic processing, with the right hemisphere specializing in spatial holistic processing. While the association between speech and the left hemisphere has been known at least since the mid-19th century [BROCA, 1861], the more extensive knowledge obtained during the past two decades has derived primarily from research on the small population of commissurotomized [GAZZANIGA, 1970; SPERRY, 1974], hemispherectomized, or lesioned patients [MILNER, 1974]. Such data cannot, of course, be readily obtained. Extension of this work depends on complementary and more accessible sources of data.

There is another, perhaps more important, reason for investing effort in the study of hemispheric specialization in the intact subject. It is becoming increasingly evident [GAZZANIGA, 1974] that while the hemispheres may well be subserving specialized functions, the organism has a capacity to integrate them in the service of whichever task is undertaken. There is a danger when focusing on abnormalities that the differentiation of neural subsystems may obscure the basic fact of neural function – namely that all the subsystems are coordinated to produce the smooth, efficient course of behavior. It is necessary therefore to investigate in detail functional differences between the hemispheres in intact subjects, in whom commissural integration is possible, to provide information on the manner in which these functional differences are recruited in the service of behavioral integration. At present, the most successful approach has been through the presentation of lateralized sensory inputs [KIMURA, 1961; BRYDEN, 1965] which allow, through the standard techniques of experimental psychology, an evaluation of differential processing [DIMOND and BEAUMONT, 1974]. Lateralization of sensory inputs, however, is not an easy task and imposes numerous restrictions on the range of paradigms in which hemispheric specialization can be studied. It is in this context that electrophysiological techniques are of potential value. If, indeed, it is possible to infer hemispheric utilization from electrophysiological parameters, then a convenient, noninvasive technique will be available to complement the data obtained from commissurotomized patients.

One promising paradigm involves the comparison of the distribution of spectral power in the alpha band at homologous hemispheric locations while subjects are engaged in tasks presumed to differentially engage the hemispheres. This interest derives from the earliest studies of the EEG by BERGER [1930] and ADRIAN and MATTHEWS [1934] which indicated an inverse relationship between alpha presence and 'mental effort'. The assumption is made that hemispheric involvement is indexed by differential suppression of alpha. The hemisphere more engaged in the task would exhibit less alpha activity than the idle hemisphere.

The same logic can be applied in studies which use ERP parameters as dependent variables. The ERP is emerging as a unique and powerful tool for probing physiological and psychological processes in humans [DONCHIN and LINDSLEY, 1969; REGAN, 1972; McCALLUM and KNOTT, 1973, 1976; DESMEDT, 1977b]. The ERP is not a uniform entity. The data accumulated over the past decade strongly supports the contention that the series of voltage oscillations, lasting several hundred milliseconds after the eliciting event, represents a composite of largely independent components related to successive levels of processing within the nervous system [McKAY, 1969; DONCHIN and LINDSLEY, 1969]. For example, the auditory ERP has been divided into three subsets of components with increasing latency from the eliciting events [PICTON et al., 1974]. An audible click will elicit a series of 7 wavelets in the first 10 msec following the event which can be recorded from the vertex in man [JEWETT et al., 1970]. Analogous components in the cat have been shown by BUCHWALD and HUANG [1975] to represent successive stages in the transmission of an auditory signal through the afferent pathways from the auditory nerve to the inferior colliculus. If the stimulus is a tone burst instead of a click, a sinusoidal component termed the frequency-following potential or FFP [MARSH and WORDEN, 1968] is elicited with approximately 6 msec latency from the stimulus. The FFP mimics the frequency of the stimulus [MOUSHEGIAN et al., 1973; MARSH et al., 1975]. SMITH et al. [1975] believe that the FFP recorded from the human scalp represents the activity of the inferior colliculus. It is notable that the FFP can be elicited by stimuli whose frequency content lies well within that of human speech [MOUSHEGIAN et al., 1973; MOUSHEGIAN, 1977].

A second group of 5 waves occur within 10–50 msec after a click. The study of these components has been hampered by their temporal proximity to reflexive muscle activity and, as yet, their origin is a matter of speculation. Nevertheless, such components have proved stable enough to be utilized for electrical response audiometry (ERA) [DAVIS, 1976; GOLDSTEIN, 1973].

The third set of components have latencies beyond 80 msec of the stimulus event. For the first 200 msec, the specific configuration of the ERP waveform appears to depend on the modality of the stimulus [GOFF et al., 1969]. These components are sensitive to a wide range of psychophysical variables [McKAY, 1969; REGAN, 1972], such as repetition rate, intensity, and frequency of the stimulus event, as well as psychological and state variables such as level of arousal and directed attention [HILLYARD and PICTON, 1974]. A number of ERP components with latencies beyond 200 msec are often recorded. These typically display the properties which characterize the endogenous components of the ERP [DONCHIN, 1975]. That is, they can be elicited in the absence of a stimulus and are largely independent of the modality of the eliciting stimulus and its specific physical features. They are, on the other hand, very sensitive to manipulations of task variables. Particularly strong data are available for a component with a latency of at least 300 msec. This component, called P300, is not modality-specific [SQUIRES, et al., 1977] and is not specifically labile to physical stimulus properties [JOHNSON and DONCHIN, 1976]; indeed, this component can be generated when the eliciting event is the absence of a stimulus [SUTTON et al., 1967]. This component is most sensitive to manipulations of complex psychological parameters such as task relevance and expectancy [SQUIRES et al., 1976]. An additional class of endogenous components precede the event of interest. Such components as the contingent negative variation (CNV) and readiness potential (RP) appear to be related to preparation or anticipation for or about the event [WALTER et al., 1964; KORNHUBER and DEECKE, 1965].

It is evident that the ERP represents progressive stages of analysis of the stimulus event. The earliest components appear to faithfully index the quality of the stimulus event and are stable over a range of psychological manipulations; hence their usefulness in ERA and neurological diagnosis [STARR and ACHOR, 1975]. With increasing latency from the event, the ERP components reflect more complex properties of the stimulus in a psychophysical and a psychological sense. Ultimately, the P300 component is independent of stimulus characteristics except in how they affect the psychological aspects of the event. For the sake of simplicity, we use the heuristic dichotomy of exogenous and endogenous components. Exogenous components are those evoked by the stimulus and thus reflect the quality and modality of the stimulus in their scalp distributions and waveshape. Endogenous components, on the other hand, are elicited by the stimulus event as it contacts information processing activities invoked by task demands and the psychological properties of the event.

Virtually all studies reviewed in this chapter have employed the same general paradigm. The independent variable is always defined in terms of tasks assigned to the subject, some presumably involving the right, others the left hemisphere. The dependent variable is always some parameter of the scalp-recorded EEG activity. The term *parameter* is used in the following sense. The primary data collected in all the reviewed studies consist of the raw EEG recorded in either analog or digital form. Any number of functions can be defined on these raw data. Such statistics as the mean power, the frequency spectrum, the cross-correlation function, or the ensemble average are all functions of the raw data. All estimate some parameter of the process which has generated the data. Thus, investigators have wide freedom in the choice of parameters. The specific choice they make is determined by their hypotheses on the nature of the EEG and EEG-behavior relations.

The studies can be conveniently classified into two categories according to the dependent variables utilized. In one category are all studies which focus on the ongoing EEG and in which, by and large, frequency domain parameters of the EEG are estimated. Such parameters are usually statements about the power or amplitude of the EEG integrated over some narrow or broad bandwidth. In the second category fall studies which analyze the EEG in the time domain. These are concerned with the waveforms of ERPs extracted from the EEG by signal averaging (one exception to this classification scheme is the coherence analysis of ERPs by DAVIS and WADA, 1974, this volume]. Within these two categories, the studies are classified in terms of the independent variables used by the experimenter.

Survey of Frequency Domain Studies

Early investigators of the EEG, while noting occasional hemispheric asymmetries, stressed the similarity of EEG tracings recorded from the two hemispheres [ADRIAN and MATTHEWS, 1934]. Large differences between homologous recordings with subjects at rest or during photic driving activation procedures were considered abnormal and used to localize unilateral focal disorders [AIRD and BOWDITCH, 1946; WALTER and WALTER, 1949; AIRD and ZEALEAR, 1951; KOOI et al., 1957]. Much evidence, however, has accrued in the past few decades that the alpha rhythm is rarely symmetric in amplitude or in phase [RANEY, 1939; RÉMOND et al., 1969; LISKE et al., 1966; HOOVEY et al., 1972]. At times, these asymmetries have been related to the subject's lateral preferences. The alpha rhythm in the dominant hemisphere has been found, by some, to be of lower amplitude [RANEY, 1939; CORNIL and GASTAUT, 1947]. However, LANSING and THOMAS [1964], in a study of predominantly dextral subjects, found a suppression of alpha in the right hemisphere. Others have found no consistent correlation [PLANVILLE and ANTONITIS, 1955; LISKE et al., 1966; RÉMOND et al., 1969; PROVINS and CUNLIFFE, 1972; BUTLER and GLASS, 1974a]. A relationship between inter-hemispheric EEG phase and laterality preferences has also been reported [GIANNITRAPANI and DARROW, 1963; GIANNITRAPANI et al., 1964, 1966; GIANNITRAPANI, 1967]. HOWEVER, the relationship appears to be quite complex and confusing, with the direction of the phase asynchrony changing with subject and state variables.

In part, this derives from difficulty in defining and validating a 'resting' state in which to take baseline EEG measures. Also, the wide variation in measurement and analysis techniques accounts for some of the confusion in the literature. Mostly the relationship between EEG laterality and subjects' lateral preferences is in fact quite complex. Interhemispheric alpha asynchrony has been reported to be more prevalent in subjects with less established lateral preferences, such as ambidextrals, or in those in whom lateral specialization may be weak, such as stutterers [TRAVIS and KNOTT, 1937; LINDSLEY, 1940]. Similar asynchronies have also been found in children with discrepant verbal motor development [LAIRY et al., 1969]. Amplitude asymmetries have been reported to be larger in subjects with clearly defined hand preferences [LAIRY et al., 1969; SUBIRANA, 1969]. EEG measures may, then, depend on the degree of lateral specialization in individuals rather than on its direction. That hemispheric symmetry of electrocortical parameters may index hemispheric integration gains support from the study of the photic driving response in human newborns by CROWELL et al. [1973, 1977] who proposed the existence of a developmental sequence of photic driving. In the earliest stage, no driving can be observed. Then unilateral (primarily right hemisphere) driving appears, followed by the bilateral driving pattern observed in adults. Such considerations must be kept in mind when evaluating the use of EEG measures to index functional asymmetry in the human brain.

A number of recent studies have claimed that interhemispheric changes in alpha and total EEG accompany the performance of functionally asymmetric tasks. Such investigations typically employ a paradigm in which a subject performs a task thought to engage primarily one hemisphere while bilateral samples of EEG are taken. Many different recording montages have been used. Occipital [MORGAN et al., 1971, 1974; DUMAS and MORGAN, 1975], temporal and parietal [McKEE et al., 1973; GALIN and ORNSTEIN, 1972; DOYLE et al., 1974; ROBBINS and McADAM, 1974], and parietal and central [RUOFF et al., 1976] electrode placements, referenced to the vertex (Cz) position have been used. Interhemispheric [BUTLER and GLASS, 1974a] and intrahemispheric [WARREN et al., 1976] bipolar linkages have also been employed. In one study, bilateral parietal electrodes were referenced to linked mastoids [CHARTOCK et al., 1975].

Some of the tasks presumed to utilize the left hemisphere differentially have included composing letters [GALIN and ORNSTEIN, 1972; DOYLE et al., 1974], word search tasks [McKEE et al., 1973], mental arithmetic [MORGAN et al., 1971, 1974; DUMAS and MORGAN, 1975; BUTLER and GLASS, 1974a; ROUFF et al., 1976], verbal listening [MORGAN et al., 1971; DUMAS and MORGAN, 1975], and verbal imagery tasks [ROBBINS and McADAM, 1974; MORGAN et al., 1971, 1974; DUMAS and MORGAN, 1975]. Right hemisphere tasks have included spatial manipulation of blocks, seashore tonal memory, and drawing [GALIN and ORNSTEIN, 1972; DOYLE et al., 1974; ROUFF et al., 1976], spatial imagery [MORGAN et al., 1971, 1974; ROBBINS and McADAM, 1974; DUMAS and MORGAN, 1975], music listening tests [McKEE et al., 1973; MORGAN et al., 1971], and affective imagery [DAVIDSON and SCHWARTZ, 1976]. In addition, occupation (artist versus engineer) and hypnotic susceptibility have been used as independent variables [MORGAN et al., 1971, 1974; DUMAS and MORGAN, 1975].

The data were analyzed in many different ways. Often, investigators integrated the raw or filtered EEG [MORGAN et al., 1971; GALIN and ORNSTEIN, 1972; McKEE et al., 1973; ROBBINS and McADAM, 1974; DUMAS and MORGAN, 1975]. Others computed amplitude histograms of the EEG [BUTLER and GLASS, 1974a] or used conventional spectral analysis techniques [DOYLE et al., 1974; ROUFF et al., 1976]. Still others have obtained measures of alpha duration over time [CHARTOCK et al., 1975; WARREN et al., 1976]. Most researchers have expressed their results in some transformation of right/left or left/right ratios for homologous electrode sites [GALIN and ORNSTEIN, 1972; McKEE et al., 1973; DOYLE et al., 1974; ROBBINS and McADAM, 1974; DAVIDSON and SCHWARTZ, 1976; ROUFF et al., 1976], or as a 'laterality score' expressing differences in power as a function of total power [MORGAN et al., 1974; DUMAS and MORGAN, 1975; WARREN et al., 1976]. Changes in these ratios are interpreted as evidence for differential hemispheric involvement. For example, GALIN and ORNSTEIN [1972] obtained the power of the total EEG at the right and left parietal electrodes. The right/left power ratio is 1.15 for the 'spatial' Kohs blocks task and 1.30 for the 'verbal' letter-writing task. The increase in the power in the right hemisphere relative to the left hemisphere for the letter-writing task is presumed to reflect the greater involvement of the left hemisphere in that task (recall that increased power implies increased alpha activity and by inference less hemispheric involvement). Similar results were obtained in studies of activity in the alpha band [MORGAN et al., 1971, 1974; MCKEE et al., 1973; DUMAS and MORGAN, 1975]. BUTLER and GLASS [1974a] found left hemisphere suppression of alpha during mental arithmetic but only in their dextral subjects; unfortunately, no right hemisphere task was used for comparison. A more sophisticated frequency analysis [DOYLE et al., 1974] revealed that the main locus of task-dependent distributional changes was in the alpha band. They reported minor hemispheric differences in the beta and theta bands and no changes in the delta band.

An interesting variant of the basic paradigm reviewed above has been used by SCHWARTZ et al. [1976] who trained subjects to produce interhemispheric asymmetries using biofeedback techniques and assessed the cognitive strategies employed by their subjects to create them: a right greater than left asymmetry in alpha was associated with more verbal cognitions while the reverse pattern was associated with more visual cognitions. These results are reminiscent of the early reports that persistent alpha production was associated with verbal imagery and alpha suppression with visual imagery [GOLLA et al., 1943].

Measures of ongoing EEG other than alpha magnitude have been used as dependent variables during the execution of functionally asymmetric tasks. One such measure is the cortical coupling index developed by CALLAWAY and HARRIS [1974] and extended by YAGI et al. [1976]. The technique attempts to assess the degree of correlation or 'coupling' between pairs of electrodes, assuming that changes in such measures represent the degree of intracortical functional communication. For related material, see LIVANOV et al. [1964]. Interhemispheric comparisons of coupling measures derived from intra-hemispheric pairs showed tasks requiring verbal-propositional processing (such as reading or mentally composing a letter) were reported to increase relative left hemisphere coupling and visual-appositional tasks (listening to music, examining pictures) to increase relative right hemisphere coupling [CALLAWAY and HARRIS, 1974; YINGLING, this volume].

Another dependent variable examined is the period of the EEG using the central-moments technique applied by SURWILLO [1975, 1976] to verbal (digit recall) and nonverbal (simple reaction time) tasks; he found no differences in central tendency, dispersion, skewness, and kurtosis in the distribution of EEG period between the left and right hemisphere (parietal-occipital derivations) during the verbal task. During the reaction time (RT) task, however, a small (0.4 Hz) but significant reduction in mean period occurred over the left hemisphere.

While these studies may indicate that there are small task-dependent changes in the EEG spectrum, the implication that selective suppression in the alpha rhythm of the dominant hemisphere for the task is the cause of the ratio changes cannot be supported on the evidence presented (unless accompanied by raw measures from each hemisphere as in ROBBINS and McADAM [1974]). It is not possible to tell if a ratio has been modified by changing the numerator, the denominator, or both when only the ratio figure has been presented. Note also that in most of these studies the experimentally induced differences are superimposed upon a constant right/left hemisphere asymmetry and do not represent shifts from a symmetric baseline. Equal increments or decrements to both the numerator and denominator of an asymmetric ratio is manifested as a change in the ratio. This point is especially important as the amount of power present may relate to the load imposed on the subject by each task.

This last point is illustrated by data collected in our laboratory. Spectral power measures were taken from subjects engaged in a visual matching task [McCARTHY and DONCHIN, 1976; an earlier version of this study based on five different subjects was reported by DONCHIN et al., 1977]. Visual stimuli, consisting of three line drawings each, were presented for short durations (50 msec) to ten subjects (all dextral without sinistral relatives). Two of the figures looked alike (structurally matched), while two of the figures were conceptually related (functionally matched). One figure was common to both pairs. The subjects were instructed to match figures on the basis of structure or function. On different experimental trials, one or the other match was required depending on the instructions. The subjects reported which of the three figures matched by pressing a single button coded for their positions (three possible responses for each slide). The reaction time as well as the accuracy of the response was measured. A tone preceded each slide by 1,000 msec. In 'fixed matching' conditions, this tone was always of one frequency and served only to warn of the slide's arrival – the subject making the same class of matches (structural or functional) for every trial. In 'mixed matching', the tone frequency randomly varied between two possibilities; each cued a different match type. During mixed mode, the subject did not know which of the two match types would be required for each individual trial until he heard the warning tone. Additional trials in which the subjects were required to respond rapidly to a neutral slide stimulus (RT control) and also trials in which no stimuli were presented at all (rest control) were used.

The performance data clearly indicated that structural and functional matching differed in difficulty regardless of the mode of presentation. Structural matching was completed more rapidly and with fewer errors than functional matching (950 versus 1,469 msec for mean RT, 93 versus 75% accuracy for structural and functional matching, respectively). The EEG spectra were measured over an epoch lasting 2,200 msec, beginning at the warning tone and lasting until 1,200 msec following the slide. Power measures were obtained for six lateral electrode sites (F3, F4, C3, C4, P3, P4) referenced to linked mastoids. In figure 1, the power in the alpha band is shown for structural and functional matching as well as for the RT and rest control conditions. The measures for structural and functional matching are pooled over fixed and mixed mode as the mode of presentation had no effect on the data.

Match type, however, had significant effects on the absolute levels of alpha. Less alpha was associated with functional than with structural matching. In addition, both control conditions were associated with more alpha than either matching task. No task dependent lateralities in alpha power were observed, implying that task difficulty was the significant factor for these data.

The generality of this interpretation might be questioned as the design of our experiment differs from others reviewed here and because the epoch analyzed for power changes included a 1,000-msec preparatory interval. However, our interpretation is strengthened by the data of CHARTOCK et al., [1975] who examined alpha power at P3 and P4 (referenced to linked mastoids) for verbal and visuospatial paired-associates tasks. These investigators found no task specific lateral differences in alpha power although, as in our study, the total alpha power present varied with some behavioral performance measure; in this case, the rate of learning for the tasks.

These data argue strongly that measures of performance should be obtained to evaluate and quantify the load the tasks impose upon the sub-

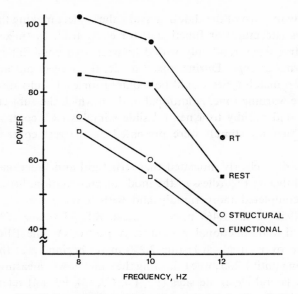

Fig. 1. Spectral power (in arbitrary units) is plotted for the alpha band for each experimental task. The values for structural and functional match conditions have been pooled over fixed and mixed presentation mode as the mode had no significant effect on alpha magnitude. The main effect of match type is significant ($p < 0.05$) at each frequency plotted.

jects. As these effects may be present, ratio measures are not valid dependent variables for evaluating hemispheric shifts in alpha power. It is interesting to note that both McCARTHY and DONCHIN [1976] and CHARTOCK *et al.* [1975] used a reference not on the scalp for recording the EEG. It is regrettable that in most of the studies presented in this section, the vertex electrode was used as the reference. Such a choice of reference clearly confounds alpha amplitude with interhemispheric phase relationships.

Survey of Time Domain Studies

In this section, we report on studies of ERPs extracted by signal averaging from the ongoing EEG. Many of the studies compared ERPs recorded at homologous hemispheric sites although some investigated only the vertex or midline positions, in order to find out whether task variables determine the relative amplitude and latency of the ERP components at the hemispheres. These differences were sometimes evaluated in terms of subjects' handedness and cerebral dominance.

1. Studies of Exogenous Components

Very few of the studies reviewed in this section were motivated by an interest in the linguistic-nonlinguistic distinction or in hemispheric specialization. Rather, the investigators were seeking information on the scalp distribution of sensory evoked potentials. Their goal has usually been the elucidation of the intracranial sources of these exogenous components. Yet often the data were collected from homologous hemispheric sites. These provide valuable baseline data on hemispheric asymmetries. Clearly, if ERPs associated with a given modality are asymmetric in the absence of any task inducement for lateralization, such biases must be considered when testing hypotheses about hemispheric specialization and the effect of linguistic or semantic analyses. The results on hand, however, are equivocal. It would be difficult to develop, on the basis of the available literature, a specification of the lateralization biases for different stimulus modalities.

The data are rather scant for the somatosensory ERPs, although it is agreed that they are largest at the scalp electrodes overlying the parietal cortex contralateral to the stimulation [GOFF et al., 1962; DESMEDT, 1971; CALMES and CRACCO, 1971; DESMEDT et al., 1976, 1977]. For the auditory modality, considerable controversy exists regarding the lateral distribution of the various components of auditory ERPs. The maximal contralateral projection to the auditory cortex as well as the oft observed dominance of one ear over the other in dichotic listening tasks suggest that, at least under certain conditions, different auditory ERPs should be recorded over the two hemispheres. Most investigators concur that right and left ear stimulation generate different scalp distributions, but there is no agreement on the specifics of these distributions. The majority of reports maintain that there is a general predominance of the contralateral response, either as a shorter latency [MAJKOWSKI et al., 1971] or as larger amplitude response [PRICE et al., 1966; VAUGHAN and RITTER, 1970; RUHM, 1971; PERONNET et al., 1974; ANDREASSI et al., 1975a], and a few in both of these measures of contralateral response [BUTLER et al., 1969]. VAUGHAN and RITTER [1970] reported a small but consistent tendency for larger responses to appear contralateral to the stimulated ear, but the effect was greater over the left hemisphere in response to right ear stimulation. Other researchers [RUHM, 1971; PERONNET et al., 1974] reported that the right hemisphere response is consistently larger only for left ear stimulation. PETERS and MENDEL [1974] failed to find such a consistent relationship between the ear stimulated and the latency and amplitude of ERP components with latency shorter than 70 msec. Given these contradictions, there seems to be little basis yet in trying to relate the lateral asymmetry of auditory ERPs to handedness, cerebral dominance or ear preference.

Similar inconsistencies appear in studies of the laterality of visual ERPs. Studies of hemispheric differences in visual ERPs have been particularly hampered by the need to assure that the ERP elicited by stimulation of a retinal half field is generated entirely within a single hemisphere. To direct stimuli at a lateral hemi-retina, the subject's fixation must be known. It is, of course, possible to instruct subjects to fixate, but it is quite difficult to be sure that they do so. The electrooculogram (EOG) can be used as a check but, without DC recording, it does not assure that the subject is in fact fixating such a point. Behavioral techniques for monitoring fixation are available. KRYNICKI [1976] provides a fine example. Whereas it has been well established that stimulation of different visual half fields elicits different scalp distribution [McKAY, 1969; REGAN, 1972; DESMEDT, 1977a], the comparison of the hemispheric distributions of visual ERPs is not as straightforward. Several investigators [KOOI et al., 1965; VAUGHAN et al., 1963; HARMONY et al., 1973] have reported that visual ERPs recorded over homologous regions in normal subjects are symmetric. Others, however, have maintained that visual ERPs recorded from the right hemisphere are larger than those recorded from the left hemisphere [PERRY and CHILDERS, 1969; RHODES et al., 1969; SCHENKENBERG and DUSTMAN, 1970; RICHLIN et al., 1971, 1976; VELLA et al., 1972; RHODES et al., 1975]. BECK et al. [1975] report that the increased amplitude of the response of the right hemisphere following blank flashes is absent (or much reduced) in sinistrals, mongoloids, and children with low IQs. This may allow for the use of the asymmetry in the visual ERP as a metric for neurological organization and development. Latency asymmetries in the ERP have also been observed. ANDREASSI et al. [1975b], for example, reported observing latency asymmetries in ERP components as a function of retinal site of stimulation.

The few investigations [EASON et al., 1967; CULVER et al., 1970; GOTT and BOYARSKY, 1972; BECK et al., 1975] concerned with the relations between handedness, cerebral dominance, eye dominance, and visual ERPs have yielded ambiguous results. EASON et al. [1967] originally reported that visual ERPs were larger over the right than the left hemisphere for left-handers only. However, a subsequent report from the same laboratory [CULVER et al., 1970] failed to confirm this finding. Rather, CULVER et al. reported visual ERP amplitudes were larger over the right than the left occipital lobe in response to left but not right visual field stimulation. This failure to replicate previous results is attributed by CULVER to confounding sex with handedness. GOTT and BOYARSKY [1972] reported that left-handers produced larger visual ERPs in the left hemisphere and that direct stimulation of the dominant hemisphere elicited ERPs with shorter latency than that elicited by stimulation of the opposite hemisphere.

2. Studies of Endogenous Components

a) Asymmetries in movement-related potentials. Little is known about the manner in which the left hemisphere is involved in linguistic function. There is, however, an aspect of brain function which is not only patently lateralized but for which the nature of lateralization is fairly well understood. This is the control of motor function. It is well known that the movements of the left side are 'controlled' from the right hemisphere and vice versa. Clearly, if hemispheric utilization has electrocortical manifestations, these should appear in association with motor function. The degree of lateral asymmetry in this context can serve as a measure of the base level of laterality. It is therefore of interest that the most consistent observations of functionally interpretable lateralization have been obtained for slow potentials which are apparently associated with the control or the monitoring of movement.

There is now a consensus that the slow negative shift preceding voluntary arm and hand movements, variously called the RP, Bereitschaftspotential (BSP), or N1 of the motor potential (MP), is a few microvolts larger over the pre-Rolandic area on the scalp contralateral to the responding limb [GILDEN *et al.*, 1966; VAUGHAN *et al.*, 1968; KUTAS and DONCHIN, 1974]. KORNHUBER and DEECKE [1965]. DEECKE *et al.* [1969, 1976], and DEECKE and KORNHUBER [1977] maintain that this contralateral dominance is restricted to the abrupt negativity just preceding the movement, but KUTAS and DONCHIN [1977] demonstrated that the hemispheric asymmetry can be observed hundreds of milliseconds prior to the response. The exact timing of the components of the MP immediately preceding the movement is, however, controversial. GERBRANDT *et al.* [1973] claimed this negativity to be post-movement while VAUGHAN *et al.* [1968] found that the MP has a somatotopic distribution and view the potentials as clearly pre-movement. Two reports [GERBRANDT *et al.*, 1973; WILKE and LANSING, 1973] reject the notion that these pre-movement potentials are associated with a motor command, and claim that these potentials are manifestations of the activity of post-response proprioceptive mechanisms. However this issue is resolved there is no question that N1 precedes the movement. Thus, our demonstration that the N1 component of the MP is larger contralateral to the responding hand is a clear illustration of the manner in which EEG scalp distributions reflect hemispheric utilization [KUTAS and DONCHIN, 1974, 1977].

These observations have recently been extended in a detailed examination by KUTAS [in preparation] of the lateral asymmetry of response-related potentials under different response conditions. In figure 2 are shown some of the results. Each pair of traces is the difference between C3' and C4' recorded when the subject responded with his right (solid line) or his left hand (dashed line) in one of the five experimental conditions described in the legend to figure 2. Increasing amplitude of these difference curves is a measure of hemispheric asymmetry, while the degree to which the curves in a pair are out of phase reflects the extent to which the direction of asymmetry depends upon the responding hand. Under all response regimes, a degree of response-related asymmetry is clearly evident, its initiation seemingly dependent upon the time at which the response selection becomes definite. The implication of these data to the study of electrocortical manifestations of linguistic function is that when clear differences in hemispheric utilization exist, they are measurable. Possibly, the relation between processing mode and the hemispheres is somewhat less definite than that between the hemispheres and motor activities.

The absolute amplitude of the MPs depends on a number of variables such as force [WILKE and LANSING, 1973; KUTAS and DONCHIN, 1974, 1977] and motivation [McADAM and SEALES, 1969]. The relevant parameters affecting the degree of N1 asymmetry, other than subject handedness and responding hand, have yet to be determined. A promising source of data is intracerebral recording from human patients [McCALLUM and PAPAKOSTOPOLOUS, 1974]. These preliminary data suggest that subtle changes in timing and asymmetry of the RP are obscured in scalp recordings.

Many investigators have noted the similarity of the CNV and N1. The suggestion that these two waveforms might represent identical processes is derived partly from the fact that most CNV studies have required a motor response to the imperative stimulus. Early mapping studies [Low et al., 1966; COHEN, 1969] demonstrated that the CNV preceding a motor response in an RT paradigm was symmetrically distributed over the two hemispheres. Within the past few years, it has been asserted that slightly larger CNVs appear over the hemisphere contralateral to the hand used for the response [SYNDULKO, 1969, 1972; OTTO and LEIFER, 1973]. SYNDULKO [1972] and SYNDULKO and LINDSLEY [1977] found that response-related lateral asymmetry was specific to central as opposed to frontal, parietal or occipital

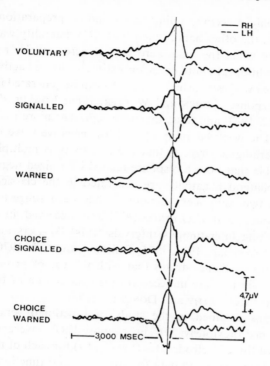

Fig. 2. Each curve represents the difference between the ERPs recorded at C3′ and C4′ calculated for each subject and averaged over the subjects for the five experimental conditions: (1) voluntary – subjects made self-paced movements; (2) signalled – subjects squeezed a dynamometer in response to a randomly occurring tone; (3) warned – subjects responded to the second of two tones 1,000 msec apart; (4) choice signalled – subjects made a right or left hand response cued by the frequency of an irregularly occurring tone; (5) choice warned – subjects responded to the second of two tones with the right or left hand depending upon the frequency of the second tone. All waveforms shown in this figure were obtained by point-to-point subtraction of the ERPs recorded at the left central electrode (C3′, 4 cm lateral to midline) from the ERPs recorded at the right central electrode (C4′, 4 cm lateral to midline). The response-locked averages obtained when subjects were squeezing a dynamometer with the right (solid line) and left hands (dashed line) are superimposed for each of the five experimental conditions. Each waveform (over 10 SS) represents an average of approximately 800–1,000 trials. The vertical line separates pre- from post-squeeze activity.

locations and developed only preceding unimanual response preparation. OTTO and LEIFER [1973], on the other hand, noted that CNV laterality was statistically significant only when the data were pooled across their 'response' and 'feedback' conditions. In both cases, only stimulus-locked activity was examined. It has been well established that CNVs can be generated in the absence of a motor response [COHEN and WALTER, 1966; LOW et al., 1966; DONCHIN et al., 1972, 1973] and must therefore represent more than mere motor preparation. The weak laterality of the slow negative wave in response-oriented CNV paradigms suggests that the negativity is multiply determined. It is conceivable that both a response-related lateralized negativity and a 'cognitive' bilateral negativity are generated in the classical CNV paradigms. Such a two-component hypothesis has been suggested [HILLYARD, 1973; GAZZANIGA and HILLYARD, 1973] and observed in a number of studies using long interstimulus intervals (ISIs) [WEERTS and LANG, 1973; LOVELESS and SANFORD, 1974, 1975; ROHRBAUGH et al., 1976]. Recent work in our laboratory demonstrates that with the use of proper distributional techniques, long ISIs are unnecessary for dissociation of the different CNV components [McCARTHY and DONCHIN, 1976].

Figure 3 presents data obtained in the study of functional versus structural matching described previously. We superimpose ERPs (averaged over all subjects) at each of three electrode sites (Fz, Cz, Pz) for each of the experimental conditions. Thus, there are data for structural and functional matches in fixed and mixed modes, as well as for the simple reaction time condition. It will be noted that the potentials recorded immediately after the tone have a different scalp distribution than those just preceding the slide. As we reported previously [DONCHIN et al., 1977], there is a clear dissociation between these two subepochs even with a warning period as short as 1,000 msec. Moreover, a principal components analysis of these data revealed the existence of two components within the epoch; one affected mostly by the presentation mode, the other largely by the match type. A more detailed description of these data will be presented by McCARTHY and DONCHIN [in preparation].

Another consideration which is often ignored in evaluation of ERP data is that when the experimental task induces much variability in the subject's reaction time, there may be a substantial difference between response-locked and stimulus-locked averages. In fact, preliminary results in our laboratory suggest that a lateralization evident in response-locked averages is obscured in stimulus-locked averages, especially in subjects with slow and highly variable reaction times.

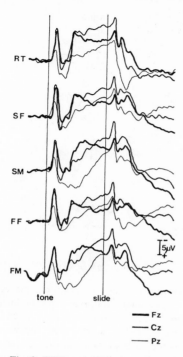

RT

SF

SM

FF

FM

tone   slide

━━ Fz
━ Cz
─ Pz

Fig. 3. ERP waveforms averaged over subjects are displayed for the RT control condition (RT), structural fixed mode (SF), structural mixed mode (SM), functional fixed mode (FF), and functional mixed mode (FM) matching conditions. Waveforms from Fz (thick line), Cz (medium line), and Pz (thin line) are overlapped. The tone preceded the slide by 1,000 msec.

Figure 4 presents individual subjects' average ERP waveforms recorded at C3' and C4' during a simple warned reaction in which the subject squeezed a dynamometer with the right hand. A comparison of the response-locked (middle column) and stimulus-locked averages (right column) reveals that the similarity between the two waveforms depends on the variability in reaction time. Response-related laterality is visible in stimulus-locked CNVs associated with low RT variability. No such laterality is observed in the stimulus-locked CNVs associated with high RT variability. The laterality is evident in response-locked averages of the same data. In another study [DONCHIN *et al.*, 1974, 1977], we have been able to elicit in rapid succession a lateralized MP and a bilateral symmetric anticipatory potential.

VOLUNTARY   WARNED   WARNED

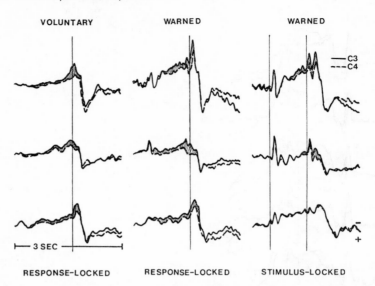

RESPONSE–LOCKED   RESPONSE–LOCKED   STIMULUS–LOCKED

Fig. 4. A comparison of event-related potentials recorded at electrodes placed at left central (C3, solid line) and right central (C4, dashed line) loci during two experimental conditions. The ERPs in columns 1 and 2 are response-locked averages obtained when subjects were squeezing a dynamometer with the right hand in a self-paced and warned situation, respectively. The vertical line represents the onset of dynamometer movement. The ERPs in the third column are stimulus-locked versions of the data shown in column 2. The two vertical lines delineate the 1,000-msec warning interval. The mean reaction times for the three SS in the warned condition from top to bottom are: 206 (SD = 42); 336 (SD = 119), and 350 (SD = 130).

b) ERP distributions and asymmetries associated with cognitive functions. Very few studies have been designed specifically to seek ERP concomitants of lateralized perceptual or cognitive functioning with the use of such endogenous components as P300 and CNV. It has been claimed that the lateral distribution of the CNV changes with task demands, but there is no consensus as to whether the engaged hemisphere has the larger or smaller CNV. MARSH and THOMPSON [1973] observed a symmetric CNV during preparation for a visuospatial discrimination, presumably a right hemisphere task. When this nonverbal task was randomly interspersed among verbal stimuli and required a pointing, rather than a verbal response, the hemisphere primary for that task had the smaller amplitude CNV. In contrast, BUTLER and

GLASS [1974b] found a larger CNV over the 'dominant' hemisphere during a warning interval in which subjects awaited numerical information. The CNV asymmetries took the form of an earlier onset and greater amplitude potential over the hemisphere contralateral to the preferred hand. Unfortunately, they only had one left-hander against which to compare the data of their right-handed subjects. Large asymmetric CNVs were also generated in the 'control' condition. Care must always be exercised in the choice of stimulus modalities and response requirements in designing these studies as CNVs in different paradigms may have different anterior-posterior scalp distributions [WEINBERG and PAPAKOSTOPOLOS, 1975]; a central dominant CNV preceding tasks requiring motor readiness [JARVILEHTO and FRUHSTOR-FER, 1970; SYNDULKO, 1972; POON et al., 1974], a frontal dominant CNV accompanying auditory discrimination [JARVILEHTO and FRUHSTORFER, 1972; SYNDULKO, 1972], and a parietal dominant CNV accompanying similar visual tasks [SYNDULKO, 1972; COHEN, 1973]. No definite conclusions can be drawn at this time as to how CNV distribution is related to cerebral dominance.

In summary, a start has been made toward utilizing ERP methods to investigate differences between the hemispheres, but progress has been slow and somewhat hampered by inadequate experimental design and analysis procedures.

c) ERPs in language reception: visual modality. Results based on multiple electrode recordings have led to the claim that asymmetric cerebral functions underlying evaluation of visual stimuli and processes related to linguistic analysis are reflected in the ERP. BUCHSBAUM and FEDIO [1969] presented different visual stimuli (words, dots or designs) in a random sequence. They reported that ERPs elicited by words can be differentiated from ERPs elicited by nonlinguistic patterned stimuli. They also claimed that verbal and nonverbal stimuli presented foveally elicit waveforms which are more differentiable when recorded at the left than at the right hemisphere. They reported similar results in a study investigating interhemispheric differences in ERPs related to the perception of verbal and nonverbal stimuli flashed to the left or right visual fields [BUCHSBAUM and FEDIO, 1970].

MARSH and THOMPSON [1973] investigated the possibility that verbal sets would lead to differential right and left hemisphere amplitudes of slow negative shifts by asking subjects to verbally identify their stimuli. During the anticipation of flashed words, symmetric CNVs were generated at the 'midtemporal' and 'angular gyrus' placements. Preliminary data obtained when the two experimental conditions (verbal and nonverbal) were intermixed yielded asymmetries in the temporal and parietal sites. Other studies dealing with visually presented words have noted a striking lack of hemispheric asymmetry. SHELBURNE [1972, 1973] recorded visual evoked potentials to three individually flashed letters which comprised either a real or a nonsense word. A comparison of the responses elicited by these two different linguistic stimuli revealed no consistent differences between the visual ERPs to the words and to the nonsense syllables in either the left or right, parietal or occipital leads. SHELBURNE's study is designed quite differently from the other studies reviewed in this section. SHELBURNE utilized a paradigm in which the last of a triplet of letters presented to a subject delivered task relevant information. This is typical of paradigms in which P300 is elicited and the waveforms published by SHELBURNE are indeed characterized by a P300 component. What distinguishes SHELBURNE's data from other P300 experiments is that the task in which P300 is elicited can be performed only by linguistically competent subjects. Thus, the extent to which the P300 is elicited can serve as an index of linguistic competence, even if no laterality is observed. In a similar paradigm, in which subjects were asked to report the key word in a visually presented sentence, no asymmetries in any of the components of the ERPs associated with words could be seen [FRIEDMAN *et al.,* 1975, this volume].

An interesting study has been conducted by KRYNICKI [1976]. What singles out KRYNICKI's study is that he preceded his electrophysiological investigation with a detailed behavioral analysis of his tasks in intact persons. Subjects were presented with 16-sided randomly shaped polygons. After each presentation, the subject had to identify the polygon by pointing to one of several in a response panel. The stimulus polygons were presented to either half of the visual field; extraordinary care was taken to assure that the subjects fixated properly. The hemispheres were shown to be differentially engaged by one aspect of the task, namely the orientation of the polygons displayed on the response panel. When the stimulus and the response figure were of the same orientation, stimulating the left hemisphere led to improved task performance; the right hemisphere was superior when the response figures were rotated. Thus, when the subject was required to perform a mental rotation of the stimulus figure to match it with a response figure, the task was performed better when the figure was presented to the left half field. KRYNICKI provides ample behavioral data in support of this assertion. The corresponding electrophysiological data are striking. The ERP elicited by the stimulus figure in the preferred hemisphere was large and elaborate. Several distinct components were clearly apparent in the right or left hemisphere ERP, depending on the subject's task. The ERPs recorded simultaneously from the other hemisphere were dramatically different in that they were characterized by a large slow component. In fact, no other study, even those making the strongest claims for electrophysiological lateralization, has reported differences of this magnitude.

An altogether different approach has been taken by CHAPMAN *et al.* [this volume] who attempt to relate aspects of processing to the waveform of the ERP. CHAPMAN's independent variable, rather than assaying a linguistic-nonlinguistic dimension, focuses on variations presumed to be inherent to language; specifically on OSGOOD's demonstration that the affective meaning of words can be described within a three-dimensional coordinate system [OSGOOD *et al.*, 1957]. CHAPMAN reports that ERPs elicited by words with extreme locations in OSGOOD's space are different than the ERPs elicited by words located in the other extreme. These intriguing results should be cautiously interpreted. The differences that CHAPMAN observed were minute (less than 1 μV); careful replication and further controls in other laboratories are clearly necessary. It is worth mentioning a potential source of experimental control that is neglected. We refer to the multiplicity of languages. It would be useful to know, for example, the electrophysiological response of bilingual speakers to CHAPMAN's word stimuli, when presented in their native and second languages. If CHAPMAN is right, there ought to be no difference between the ERPs elicited by words from the same locus in OSGOOD's space no matter in which language they were presented. Also, presenting words in a language foreign to a subject and comparing his response to that of native speakers should reveal if it is the affective meaning or the physical features of the stimuli which determines the response.

d) ERPs in language reception: auditory modality. Although no lateralized component specific to either linguistic or non-linguistic stimuli has been identified reliably, it has been amply demonstrated that such stimuli yield ERPs with distinct features when presented visually. On the other hand, investigators have been unable to differentiate consistently between ERP waveforms to aurally presented linguistic and non-linguistic stimuli. Vertex ERPs time-locked to the first syllable of sentences which had to be identified in the presence of interfering messages and under conditions of distortion were found to be very similar to tone ERPs [FELDMAN and GOLDSTEIN, 1967]. Furthermore, while the ERP waveforms in this task varied as a function of signal level, they were not influenced by sentence intelligibility. ROTH *et al.* [1970] could easily classify the responses to the beginnings of sentences as a function of task, whether sentences were attended to or ignored, but could not distinguish between the ERPs to the different syllables in the sentences. They were also unable to discriminate ERPs to real words from those to nonsense syllables. The presentation of stimuli, however, in a story form with nonsense syllables interspersed required linguistic analysis of each of the monosyllables independent of semantic value. RATLIFF and GREENBERG [1972] – employing various tones, words, and noise envelopes – obtained differences in the latency and polarity of the ERP components between linguistic and non-linguistic stimuli, being longer for nonlinguistic than linguistic stimuli. GREENBERG and METTING [1974] further noted that while normal subjects display different ERPs to speech and non-speech stimuli, aphasics generate abnormal and inconsistent responses to all stimuli.

Although still contradictory and inconsistent, more promising results relating ERP lateralization to linguistic or semantic processing have been obtained with auditory stimuli [GREENBERG and GRAHAM, 1970; COHN, 1971; MORRELL and SALAMY, 1971; WOOD et al., 1971; MATSUMIYA et al., 1972; TEYLER et al., 1973; BROWN et al., 1973, 1976; HAALAND, 1974; NEVILLE, 1974; MOLFESE et al., 1975; WOOD, 1975]. A number of studies in which various combinations of phonemes, words, and sentences were used to elicit ERPs have supported the view that linguistic analysis occurs primarily in the left hemisphere while processing of nonverbal stimuli, such as musical chords, takes place in the right hemisphere [TAUB et al., 1976]. GREENBERG and GRAHAM [1970] investigated the EEG correlates of verbal learning through an analysis of spectral components of AEPs to various consonant-vowel-consonant combinations (CVC) and piano notes. They found a left hemisphere dominance of the activity in all frequencies during speech learning and a trend toward a shift from left to right hemisphere dominance in the frequency components considered to reflect overall ERP amplitude for the notes. TAUB et al. [1976] in fact demonstrated that the averaged response evoked by monaurally presented nonverbal sounds, such as major, minor, and diminished musical chords, are larger over the right than over the left auditory cortex. In a brief report, COHN [1971] tells of a prominent, positive-going peak with a 14-msec latency elicited in the right hemisphere by click stimuli but not by single syllable words. MORRELL and SALAMY [1971] found that the N100 component elicited by nonsense words was larger over the left than right temporoparietal electrode. It is difficult to interpret their results as they failed to use a non-language control. MATSUMIYA et al. [1972] reported a hemispheric asymmetry in a 'W-wave' (a positive response recorded bipolarly, peaking at 100 msec) elicited by real words and environmental sounds. They ascribed this hemispheric asymmetry to the significance of the auditory stimuli for the subject rather than to the linguistic features of the stimulus. WOOD et al. [1971] reported differences in the ERPs recorded over the left hemisphere which appeared in the N100–P200 component, depending on whether the subject was required to perform a linguistic or an acoustic analysis of the stimulus. These results were essentially replicated by WOOD [1975, this volume] in experiments comparing the processing of two different dimensions (auditory and phonetic) of the same CV syllables. Distinct ERPs were obtained to stimuli along phonetic and auditory dimensions but not to two stimuli along the auditory dimension or different acoustic components of the phonetic dimension. MOLFESE et al. [1975] found a similar enhancement in the amplitude of the N100–P200 component of the ERP in the left relative to the right hemisphere for speech stimuli even when the subjects' task was merely to listen. On the other hand, nonspeech acoustic stimuli were found to produce larger amplitude responses in the right hemisphere. The lateral differences to both types of stimuli decreased with age from infants to adults [MOLFESE, this volume]. GALAMBOS et al. [1975] failed to find hemispheric differences in the P300 component elicited by target stimuli although they did note that the P300 to speech stimuli were of longer latency than those to clicks. NEVILLE [1974] reported lateral ERP amplitude and latency differences elicited by digits but not by clicks in a dichotic listening paradigm. During a comparison of the effects of different forms of verbal input (monaural, dichotic and diotic) on inter- and intrahemispheric ERP distributions, HAALAND [1974] found that the P(160–306) component was greater over the right than the left hemisphere in all presentation modes.

Several investigators have attempted to evaluate the influence of linguistic meaning on scalp ERPs. TEYLER et al. [1973, this volume] recorded different ERPs from the same electrode site to the same click stimulus depending on the meaning of the verbal context (noun-verb) to which the stimulus was temporally related. Linguistic stimuli elicited responses of greater magnitude in the dominant hemisphere. In a similar study, BROWN et al. [1973] recorded ERPs to the actual words rather than to coincidental clicks. The words they used were ambiguous and were disambiguated by their context. They reported (1) that the waveform of the ERPs evoked by the same word differed according to its contextual meaning, and (2) that these differences were significantly greater for left than right hemisphere loci. Similar results were obtained by BROWN et al. [1976] in a study using the stimulus homophone 'led' or 'lead' in the ambiguous phrase 'it was /led/'. The ERPs elicited by the homophone differed as a function of the subjects' set to perceive a particular meaning of the phrase only in the left hemisphere, primarily over Broca's area. It seems, then, in general, that different investigators find in a variety of ERP parameters greater variability over the left than the right hemisphere.

Two strategies are available for revealing the effects of hemispheric specialization on ERP components. Both strategies challenge the subjects with tasks presumed to engage selectively one of the hemispheres. One, however, investigates the ERPs elicited by stimuli integral to the challenging task. This has been the strategy adopted by the studies reviewed above. An alternate strategy would probe the subject with stimuli unrelated to the task. If the hemispheres are differentially engaged, the response evoked by such probe stimuli may well be different. The more direct strategy adopted by most investigators is somewhat weak because the choice of differential tasks is dictated by the requirement that they provide distinct stimuli for eliciting ERPs. The probe strategy permits the choice of tasks which are continuous in time and with a potential for substantially loading the hemispheres. The ERP to the probes during task performance could be compared to the response to the same probes in the absence of a task.

This approach has been used rather infrequently. A report by GALIN and ELLIS [1975] indicated that the symmetry of the visual ERP is influenced by the spectral characteristics of the EEG at the time of stimulus presentation. They found that ERPs elicited during tasks which induced hemispheric asymmetries in alpha power were also asymmetric as determined by measures of peak-to-trough amplitude and power. A sophisticated application of the probe strategy has been reported by SHUCARD et al. [1976], who demonstrated that lateralization of brain functioning can be indexed by task-irrelevant tone pairs which were presented while subjects were either detecting a melody in a Bach fugue or listening for a phrase in a spoken passage. They report that the probe ERPs were of higher amplitude in the hemisphere hypothesized to be most engaged by the primary task (the right for musical task and left for verbal task). The effect was most pronounced for the second of the paired tones. Unfortunately, as SHUCARD et al. [1976] recorded T4-Cz and T3-Cz, it is difficult to determine if their results are due to shifts in hemispheric balance or in the intrahemispheric distribution of the potentials. Such results are provocative and suggest that baseline symmetry in ERPs may depend upon variability in ongoing EEG activity which may, in turn, depend upon subject state variables.

e) *Slow potentials preceding speech production.* Although some questions were raised as to the replicability and exact nature of the reported ERP differences associated with speech perception, no doubts were cast as to the cerebral origin of these potentials. In contrast, the interpretation of scalp-recorded activity associated with speech production present difficulties in differentiating between cerebral potentials and interfering neuromuscular and respiratory artifacts. This possible confound has led to controversy concerning the origin, topographical distribution and hemispheric asymmetry of these potentials and their relationship to cerebral dominance.

A number of bioelectric sources of artifact such as eye movements, GSR, head movements, muscle tension, respiratory waves, and lip and tongue movements can influence asymmetrically speech-related electrical potentials [KLASS and BICKFORD, 1960; MORRELL and HUNTINGTON, 1971, 1972; GRABOW and ELLIOTT, 1973, 1974; GRÖZINGER et al., 1973, 1975, this volume]. KLASS and BICKFORD [1960] were the first to note that movements of the tongue, the tip of which is negatively charged in relation to the body, gave rise to 1–6/sec rhythmic waves in the EEG of 5% of their patient and subject samples during speaking and reading aloud. The prominence of this artifact, labelled the glossokinetic potential, appeared to be a function of '...(1) the degree of tongue movement with speech; (2) the cranial topography of the tongue relative to the scalp, and (3) the electrical charge on the tongue...'. Since then GRABOW and ELLIOTT [1973, 1974] found that hemispheric asymmetries in the EEG could be produced by directing the tongue toward either the right or left side of the mouth. GRÖZINGER et al. [1973, 1975] further noted that articulation and speech can be preceded by slow periodic brain potentials (R waves) correlated with respiration which also seem to have a lateralized distribution, the more active hemisphere showing slightly smaller potentials [GRÖZINGER et al., this volume].

Few studies, however, have acknowledged the possible presence of artifacts from lateral eye movements (not blinks) which themselves have been accepted as reflective of functional hemispheric asymmetry. Shifts to the right have been associated with left hemisphere processing while shifts to the left are taken as indicators of right hemisphere engagement [KINSBOURNE, 1973, 1974; GUR et al., 1975].

Investigating cortical activity time-locked to speech onset, SCHAFER and ERTL [SCHAFER, 1967; ERTL and SCHAFER, 1967] maintained that reliable potential changes free of muscle and movement artifact occur over the right sensorimotor and left temporal speech association areas following and preceding speech. Using himself as a subject in 15 different experiments,

SCHAFER [1967] found that these cortical command potentials time-locked to voice activation were characteristic of each of the spoken stimuli including the letters T, O, and P, the numbers two and ten, and the words yes and no. SCHAFER tentatively suggested that the early pre-speech components might correspond to the acts of word selection and the decision to speak while the later pre-speech components might reflect the initiation of pyramidal tract firing which controls the vocal musculature specific to a particular word. ERTL and SCHAFER [1967] extended these findings for the right sensorimotor area in five additional subjects during a comparison of potentials preceding speech and those preceding fist contractions. Interpretations of the cortical command potential described by ERTL and SCHAFER are limited by their use of time constants too short to allow the emergence of activity with frequency in the RP range. However, VAUGHAN et al. [1968] reported that MPs associated with facial and tongue contractions were similar in form to those obtained with limb movements. A subsequent finding that simultaneous recording from the upper lip muscles resulted in an averaged waveform of similar time course to that recorded from the scalp electrodes led ERTL and SCHAFER [1967] to suggest lip movement as a possible source of the potentials.

The first extensive study of the relationship between cortical activity and the generation of a variety of speech and non-speech sounds was carried out by MCADAM and WHITAKER [1971]. They observed a small increase in the negativity over Broca's area (in the left hemisphere) preceding spontaneous spoken words but not preceding simple oral gestures such as spitting and coughing. This report, however, has been attacked by MORRELL and HUNTINGTON [1971] on several grounds. They questioned MCADAM and WHITAKER's procedures, analyses, and conclusions. MORRELL and HUNTINGTON claimed that when the articulatory muscles were monitored and the same measurements were made for all waveforms, no hemispheric asymmetries consistent with localization over Broca's area could be found. In a subsequent study, however, MORRELL and HUNTINGTON [1972] reported that even after the contaminating effects of lip muscle activity were removed, there remained cortical potentials of cerebral origin time-locked to speech production.

Other examinations of the contamination by artifacts of brain potentials preceding speech production have led to conflicting conclusions. In another attempted replication of MCADAM and WHITAKER [1971], GRABOW and ELLIOTT [1973, 1974] found no consistent hemispheric electropotential lateralization before, during or after verbalization. GRABOW and ELLIOTT emphasized the possible contamination by the glossokinetic potential as well as the hemispheric asymmetries in the EEG which could be produced by tongue movement. They concluded that there was insufficient evidence relating the average ERP to coding for language production in either hemisphere. On the basis of topographic analysis of speech-related potentials, SZIRTES and VAUGHAN [1973] similarly concluded 'the cortical potentials associated with muscle innervation and possibly with reauditorization of speech sounds overshadow neural activity within the dominant hemisphere related to the programming of speech'.

On the other hand, a number of investigators [Low *et al.*, 1973, 1976, this volume; GRÖZINGER *et al.*, 1973, 1975, this volume] concurred with the statement that the distribution of the pre-speech cortical potentials obtained after elimination of possible artifacts is consistent with an intracranial source reflecting preparation for speech production and the known asymmetries between the hemispheres for linguistic processing. Low *et al.* [1973] have provided further support for such conclusions by finding a significant correlation between hemispheric dominance as determined by the Wada sodium amytal test and dominance derived from the relative CNV amplitudes in the left and right motor speech area. ZIMMERMAN and KNOTT [1974] applied similar procedures to an investigation of the physiological basis of stuttering. A comparison of CNVs in stutterers and normal speakers during speech and nonspeech tasks revealed that only 22% of the stutterers showed a left greater than right asymmetry as opposed to 80% of the normal speakers.

Thus, although a substantial amount of clinical data support the theory of left hemisphere superiority in language reception and production, the ERP data regarding this functional asymmetry are far from consistent. The methodological and statistical shortcomings which exist in some of the studies cited along with inconsistencies in the others render any decision about the efficacy of ERPs as indices of linguistic processing inconclusive.

Concluding Remarks

Much of the research reviewed in this chapter can be described as a search for the electrocortical manifestations of the activity of a left hemisphere language processor. The assumption is made that tasks which require the extraction of linguistic meaning from auditory or visual signals will activate a unique population of neurons charged with such an analysis.

The activation of the linguistic processor will, in turn, produce an interpretable pattern of electrical potentials at the scalp. This same logic applies to the activity of that other impalpable processor, presumably located in the right hemisphere, which is charged with spatial or wholistic processing of stimuli. The crux in both cases is the selection of a task which will activate the appropriate processor.

Unfortunately, the selection of tasks often proceeds on intuitive grounds using the most general statements concerning hemispheric specialization. A severe problem confronting investigators is the lack of a clear consensual theory which could be used to determine which tasks should have what effects on the electrophysiological measures. It is, therefore, no surprise that much of the research is descriptive. In essence, the study of language and the brain and hemispheric specialization was founded on a collection of observations of clinical cases. What theory there is has accumulated over time as the data base expanded and was systematized. Nevertheless, if electrophysiological measures are to add to the study of such complex processes of the brain (rather than merely to reaffirm current knowledge), then tasks must be chosen with reference to existing theoretical frameworks. This, combined with validating behavioral measures, insures that electrophysiological changes, if found, will illuminate rather than obfuscate.

Problems in evaluating many of the studies are compounded by the lack of consistency in the data concerning the baseline symmetry of electroencephalographic parameters. However, asymmetry in baseline conditions does not immediately imply poor electrode placement or asymmetries in subject's skull thickness. Considerable data exists [BECK et al. 1975; CROWELL et al., 1973; LAIRY et al., 1969; LINDSLEY, 1940; TRAVIS and KNOTT, 1937], suggesting that the symmetry of electrophysiological measures may vary across individuals as a function of the degree of hemispheric differentiation and its development. Experimenters are therefore faced with the necessity of accounting for a large range of subtle subject factors in their experimental designs. While it is common to treat hemispheric specialization as a polarized process, the behavioral evidence argues for gradients of differentiation in the 'normal' population [SHANKWEILER and STUDDERT-KENNEDY, 1975]. Thus, intuitive task selection based on the most general statements of hemispheric specialization will no longer bring a good return in knowledge with respect to experimental investment.

It may be more profitable to develop experimental paradigms in which ERPs are mobilized to search for answers to questions of linguistic relevance. Language is a form of information processing. There is considerable evidence that the ERP, in particular the endogenous components, are valuable indices of various aspects of human information processing. It would seem of value to structure experiments not around a search for lateralization *per se,* but on a correlation between linguistically determined aspects of information processing and ERP parameters. Consider for example the issues reviewed by WANNER *et al.* [this volume]. Of particular import is the sequential nature of sentence processing. Is a sentence perceived as a unit or is it processed sequentially? This issue may be tractable to an ERP approach. FRIEDMAN *et al.* [1975] have shown that distinct words in a sentence elicit P300 components and that their amplitudes vary with the degree to which each word is crucial to the meaning of the sentence. Several studies from this laboratory have shown that the amplitude of P300 varies with the expectancy of a stimulus and that the expectancy is determined by the fine structure of events immediately preceding that stimulus [SQUIRES *et al.,* 1976; DUNCAN-JOHNSON and DONCHIN, 1976]. This work suggests that it is possible to evaluate the sequential dependencies within linguistic structures by a study of their ERP concomitants. This is all the more possible if advanced statistical procedures, which are commensurate with the complex nature of ERP data, are used. The dissociation of ERP components with the use of principal components analysis [DONCHIN, 1966, 1969] and the ability to study ERPs associated with single events [SQUIRES and DONCHIN, 1976] allows for a variety of experimental designs.

Another useful variable is the latency of ERP components. While ERP amplitude has been the dependent variable in most ERP studies, the latency of some components, particularly the P300, can be a rich source of information about the manner in which subjects process stimuli. SQUIRES *et al.* 1977] have shown that the latency of P300 depends on the duration of stimulus evaluation. KUTAS and DONCHIN [in press] recorded the P300 elicited when the subject was required to semantically categorize words. The more complex the semantic categorization, the longer the latency of P300. Moreover, in KUTAS *et al.* [1977], a procedure is described for measuring the latency of P300 on individual trials. When this procedure is applied to the semantic categorization data, we find that the latency of P300 reflects, on a trial-by-trial basis, the time required by the subject to categorize the stimulus. Given these data, it is possible to use P300 as an indicator response in psycholinguistic studies where it can augment traditional reaction time measures [cf. also RUCHKIN and SUTTON, 1978].

Our intent is not to detract from studies searching for electrocortical manifestations of verbal and spatial processing, but rather to direct atten-

tion to the many contributions which ERP studies can make to the understanding of human information processing beyond the reaffirmation of
hemispheric specialization. We are convinced that these approaches are
complementary and necessary.

Prof. EMANUEL DONCHIN, Cognitive Psychophysiology Laboratory, Department of Psychology. University of Illinois, *Champaign, IL 61820* (USA). Tel. (217) 333 3384.

Language and Hemispheric Specialization in Man: Cerebral ERPs.
Prog. clin. Neurophysiol., vol. 3, Ed. J. E. DESMEDT, pp. 243–253 (Karger, Basel 1977)

Hemispheric Asymmetry and the Effects of Attention on Visual Evoked Potentials[1]

MONTE S. BUCHSBAUM and DOROTHY DRAGO

Adult Psychiatry Branch, National Institute of Mental Health, Bethesda, Md.

The left cerebral hemisphere has long been recognized as the major processing area for verbal information and the right hemisphere as the area for certain nonverbal, spatial information [cf. SPERRY, 1974]. It is attractive to hypothesize, as did KIMURA [1966], that subjects would perform faster and/or more reliably on perceptual tasks if the stimuli could be presented directly to the hemisphere specialized for the task. The anatomical basis for such direct information flow is well known. Stimuli from the right visual hemi-field are projected onto the left side of the retina of both eyes; from there optic fibers run through the optic chiasm crossing so that they go to the left lateral geniculate and left occipital cortex.

Recent studies have provided suggestive evidence for hemispheric asymmetry in event-related potentials to speech and sound effects, with larger EP components in the left hemisphere of right-handed subjects for speech stimuli [this volume]. Recently, in a study of hemispheric asymmetry in EPs to speech and sound effects, MATSUMIYA et al. [1972] suggested that attentional effects might explain the observed larger left hemisphere EPs for speech stimuli. They noted that left hemisphere (Wernicke's area) EPs were larger for mechanical noises as well as speech when subjects were asked to count the number of different types of noises. These suggestions by MATSUMIYA et al. [1972] and KINSBOURNE [1971] that attentional processes could show lateralization together with reports of lateralization of simple perceptual responses led us to study the relative size of the EP attentional effect in the left and right hemisphere when hemiretinal stimulus presentation is varied.

---

[1] The authors thank R. COPPOLA and C. KING for technical assistance.

Methods

29 right-handed adult female volunteers served as subjects. The stimuli were lights of two intensities, 0.5 and 50 foot-lamberts, presented as semicircles with their centers 2° to the right or left of an illuminated fixation spot. Each stimulus could be illuminated separately from behind by a fluorescent tube driven by an Iconix Photostimulator under computer control [GIPS et al., 1971]. The stimuli had a 3-msec rise time, 0.5-sec duration, and their onset was no less than 1 sec apart. Stimuli were presented in a pseudo-random sequence such that each visual field and each intensity were presented an equal number of times and were preceded by each other condition an equal number of times.

The stimulus control computer program allowed stimulus presentation only during periods when subjects fixated on the central illuminated spot. The subject was seated facing the photostimulator screen and had her head positioned firmly in a headholder chin rest to minimize head movements. The subject's fixation was monitored using a special television camera device. Acceptable fixation was defined as the area within a 10′ visual angle of the fixation light. The subject's fixation was monitored using Mackworth eye movement camera stand optics [MACKWORTH, 1967] and a television camera device [BUCHSBAUM et al., 1973]. The TV camera picked up the corneal reflection and from the video signal a special purpose analog device output a voltage proportional to the x and y coordinates of the subject's point of gaze. This was sampled continuously by the computer. So long as fixation remained on the fixation light, stimuli continued to be presented. If a subject lost fixation, the stimuli were not presented until she returned her gaze to the designated spot. The time delay between the return of the eye to correct fixation and the stimulus presentation was at least 50 msec and generally longer due to the time required for the TV raster scan, vidicon tube intensification and 50 Hz sampling rate of the computer.

The EEG was recorded between O1, O2, W1, and W2, and the ipsilateral ear, with Cz as ground. W was located over the Wernicke area, as described in MATSUMIYA et al. [1972]. The EEG was amplified (3 dB bandwith at 0.3 and 116 Hz), filtered (40 Hz cutoff, down 42 dB at 60 Hz) and sampled at 125 Hz for 500 msec following each stimulus onset. The EP average was divided into three time bands (76–112, 116–152, and 168–248 msec) based on the visual observation of the three major peaks (P100, N140, and P200) usually identified in VEPs. The mean value of the EP was subtracted from each coordinate as a baseline and the mean of the absolute values of successive points 8 msec apart was calculated within each time band [cf. SHAGASS, 1972].

Three tasks requiring attention to the visual stimuli and one task which would require ignoring the stimulus were chosen. The identical sequence of visual stimuli was presented each time. The four VEP runs were made in a random order: (1) mental arithmetic (MA) – mental subtraction of seven serially from 2,019; (2) intensity discrimination (ID) – subjects assigned mentally a value of 1 to the high intensity stimulus and 2 to a low intensity stimulus; (3) spatial pattern (SP) – subjects counted the number of times they observed two consecutive flashes on one side followed by two consecutive flashes on the other side regardless of intensity; (4) intensity pattern (IP) – subjects counted the number of times they observed two consecutive flashes of one intensity followed by two consecutive flashes of the other intensity regardless of the side of presentation. For the SP and IP task, practice sessions were held to insure that the subjects understood the task. The 'intensity' or 'spatial' patterns occurred about once every 8–10 stimulus flashes (a total of 26 times in

256 stimuli). Subjects were asked to report the number of patterns counted and nearly all subjects were correct within ± 3 of the correct number.

Results

Effects of attention. As expected, VEPs were larger when subjects attended to the visual stimuli than when they did mental arithmetic. Figure 1 shows a typical subject's VEP for the IP and MA conditions. Out of 144 of the combinations of EEG lead, EP component, hemisphere, visual field, intensity and the three attentional tasks, 123 showed larger mean VEP amplitude than VEPs recorded during mental arithmetic. The N140 (116–152-msec time period) component showed the greatest attentional effect (fig. 2). Increases in amplitude were, for the 76–112, 116–152, and 168–248-msec time bands, 35, 50, and 6%. For the 116–152-msec time band, each of the 48 conditions showed an amplitude increase with attention.

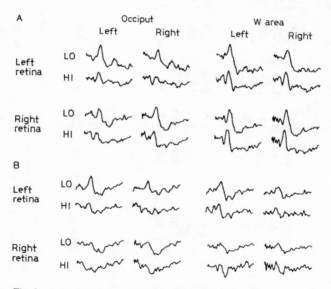

Fig. 1. A Typical subject's EPs for intensity pattern task for occipital and Wernicke's area leads for LO and HI intensity stimuli (0.5 and 50.0 foot-lamberts). Each EP is 500 msec long, and drawn with negative up. Note small size of ipsilateral EPs. *B* EPs recorded during mental arithmetic, occipital and Wernicke's area; same subject as in *A*.

Table I. Mean EP amplitude

Task	Intensity	Left hemisphere						Right hemisphere					
		occipital			Wernicke's area			occipital			Wernicke's area		
		B1	B2	B3	B1	B2	B3	B1	B2	B3	B1	B2	B3
Left retina													
IP	0.5	4.9*	8.4*	6.2	5.6*	8.5	4.6*	3.7	4.1	4.7	4.7*	5.3*	3.2
MA	0.5	4.4	5.5	5.6	3.0	3.8	3.2	3.5	3.2	3.7	2.3	2.6	2.8
IP	50	6.1	7.1*	5.2	5.1*	7.6*	4.3	4.7	5.5*	3.5	3.6*	5.7*	3.2
MA	50	5.1	5.2	5.7	3.3	3.7	4.0	4.2	4.1	4.3	2.3	3.9	3.2
Right retina													
IP	0.5	3.9*	4.7*	4.4	4.1*	3.5*	3.4	4.4	7.0*	5.3	3.7*	6.1*	3.3
MA	0.5	3.0	3.3	4.7	2.3	3.0	3.5	4.0	4.3	5.3	2.3	3.4	3.3
IP	50	4.9*	6.1*	4.7	4.3*	6.9*	4.4	5.4*	6.8*	4.3	3.6*	6.6*	3.5
MA	50	3.9	4.3	5.0	3.1	4.5	3.8	4.1	4.1	5.1	3.0	4.0	3.4
Mean attention effects													
Left retina													
IP–MA	0.5 ε 50	*0.72*	*2.41*	*0.15*	*2.17*	*4.29*	*0.82*	*0.34*	*1.15*	*0.12*	*1.86*	*2.24*	*0.21*
Right retina													
IP–MA	0.5 ε 50	*1.03*	*1.62*	*0.34*	*1.49*	*1.44*	*0.29*	*0.88*	*2.65*	*0.40*	*1.04*	*2.64*	*0.04*

* Significantly larger (p<0.05) EP amplitude for attentional task (IP) than distraction task (MA).
Italic numbers indicate direct pathway responses

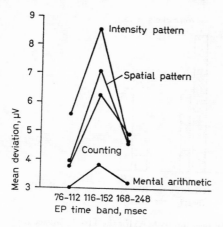

Fig. 2. Mean EP amplitude for three time bands for left hemisphere. Wernicke's area, right visual field stimuli for the four subject tasks. The tasks requiring attention to the stimulus have clearly larger EPs than those recorded while subjects did MA. Note that the 116–152-msec time (N140) band shows not only the greatest attentional effect, but that the amount of attentional effect is related to task difficulty.

The attention effect was greatest for the most difficult task, intensity pattern and least for the easiest task, intensity discrimination. Attentional effects were calculated by subtracting the MA condition VEP amplitude from each attentional task VEP amplitude; a positive number thus indicating VEP amplitude enhancement with attention. The relationship between task difficulty and VEP enhancement was seen most clearly in the N140 time period. The attentional effect for the IP, SP, and ID tasks was 1.95, 1.39, and 1.28 μV for the occipital lead, and 2.66, 2.52, and 1.87 for the W lead (averaged across field hemisphere and intensity). Data for the intensity pattern and MA task are presented in table I.

Attentional effects were greater for direct VEPs than indirect VEPs (fig. 3). Table II, under mean attention effects, shows the extent to which direct EPs showed greater attention effects than indirect EPs. Clearly, only the 116–152-msec time band showed either much attention effect or much attention asymmetry. This time band also shows increasing direct-indirect attentional differences which increased with task difficulty. The IP task also showed greatest EP attentional effect superiority for direct in comparison with indirect (table II). The 2 way ANOVA of direct-indirect differences showed a significant task effect ($p < 0.05$) for the 116–152-msec band for both occipital and Wernicke's area.

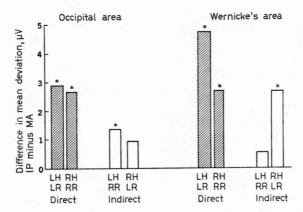

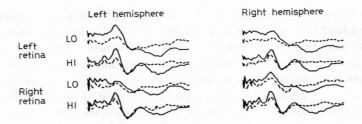

Fig. 3. Mean EP attention effect (IP task minus MA) for the 116–152-msec time band (peak N140) for each lead (occipital and Wernicke's), hemisphere (LH and RH) and each hemiretina of stimulus presentation (LR and RR). Asterisks indicate an attention effect significantly different from zero (*t*-test, p <0.05). Note that direct route EPs show larger attention effects than indirect EPs.

Fig. 4. Mean EPs calculated across all 29 subjects for Wernicke's area. IP task EPs shown as solid line and MA EPs shown as dotted line. Note EP amplitude increase for negative component at about 140 msec in IP task over the same component in the MA task. P300 enhancement is also visible.

W area leads exhibited the greatest amplitude increase with attention (66% averaging across all conditions) in comparison with occipital leads (33%); the optimal combination of conditions to maximize attentional effects (left hemiretina stimuli, left W area recording, band 2 and low intensity) produced more than a doubling of average EP amplitude size (fig. 3). This amplitude increase with attention was significantly greater than

Table II. Mean EP amplitude: direct (LHLR + RHRR) minus indirect (LHRR + RHLR) averaged across intensities

Task	Occipital			Wernicke's		
	B1	B2	B3	B1	B2	B3
IP	0.87	2.19	0.93	0.30	1.82*	0.36
SP	0.89	1.68	0.82	0.12	1.51*	0.43
ID	0.60	1.68	0.80	0.02	1.45*	0.00
MA	0.77	1.04	0.94	0.37	0.22	0.43
Mean attention effects						
IP	0.10	1.15	−0.01	−0.07	1.60	−0.07
SP	0.12	0.64	−0.12	−0.25	1.29	0.00
ID	0.17	0.64	−0.14	−0.35	1.23	−0.43

* $p < 0.05$, difference between direct and indirect greater than zero.

its right hemisphere direct pathway (RRRH) counterpart, than the RRLH attention effect, and than the LRLH occipital response (by *t*-tests, all $p < 0.05$). Figure 4 shows mean EPs calculated across all subjects and graphically illustrates the results found in the various ANOVA. Figure 5 shows the 40-msec segment (116-152 msec) quantified as N140 for all 29 subjects for attention (solid) and MA (dotted). This illustrates that while most subjects show a distinct negative peak at this time interval (especially for LHLR), considerable diversity exists as to waveform. For this reason, latency measurement or peak-to-trough hand measurement is difficult and we have used the area integration technique for EP measurement.

Direct-indirect comparisons. Direct pathway EPs (left-retina stimulation-left hemisphere EP plus right-retina stimulation-right hemisphere EP) were significantly larger than indirect pathway EPs (right retina-left hemisphere plus left retina-right hemisphere) for nearly every time band and lead (table II). Direct pathway superiority was greatest for low intensity stimuli. This was confirmed statistically with a 2-way analysis of variance on the direct-indirect difference scores. With task and intensity as the two dimensions, the intensity effect was significant for the 76–112 and 116–152-msec time bands for both occipital and Wernicke's lead. For the positive peaks, direct-indirect differences were greatest in the occipital leads (table II).

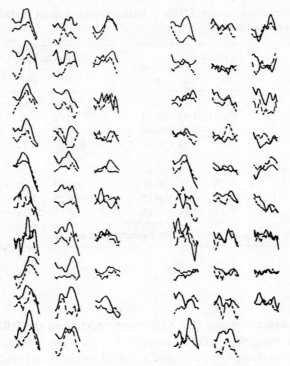

Fig. 5. Individual EP segments from 116–152-msec time band for all 29 subjects for Wernicke's area, low intensity. EPs for IP task are shown as a solid line. MA EPs as a dotted line. EPs are plotted relative to their own mean level. Left: Direct (left hemisphere, left retina) responses. Subjects are arranged in 3 columns, grouped roughly according to gross EP waveform. The first column has primarily sharply pointed negative peaks, the second broader traces and the third irregular or flat configurations. Most subjects, at least for left hemisphere, left retina EP, show a distinct negative peak in this time period. Right: Indirect (left hemisphere, right retina) EPs for subjects in same order. Note that the negative peak is much less clear and that the difference between IP and MA EPs is much diminished, especially in persons in the first two columns.

Contralateral VEP. Left hemisphere VEPs were larger than right for direct route comparisons at low intensity (*t*-test, $p < 0.05$, occipital and W area means 116–152 msec) for the IP, SP, and ID tasks; differences were smaller at high intensity. As with direct-indirect differences, direct left-right differences were greatest with the most difficult task and least with MA. The low intensity LHLR minus RHRR differences for 116–152-msec time band for IP, SP, ID, and MA were 2.50, 1.68, 0.76, and 0.49 μV, respectively. For indirect route comparisons (LRRH and RRLH), the right hemisphere responses were greater, again only at low intensity. Hemispheric differences followed task difficulty. The low intensity LHRR minus RHLR differences for 116–152-msec time band for IP, SP, ID, and MA were −0.74, −0.64, 0.69, and 0.41 μV, respectively. Since the appearance of a VEP in the hemisphere opposite to the hemiretina stimulated depends at least partly on interhemispheric communication, larger indirect VEP could suggest selective transfer of information to the right hemisphere. The MA task, requiring no such transfer, showed no such left-right asymmetry. It should also be noted that differences between hemispheres were relatively smaller than either attentional effects or direct-indirect route differences.

Discussion

Both attention and language effects most frequently involved component N140 (timeband 116–152 msec). The finding of our greatest attention effect for this negative component is consistent with many reports on event-related potentials [cf. this series]. Language effects also seem maximal in the 116–152-msec region. MORRELL and SALAMY [1971], using speech stimuli, found the greatest hemispheric differences in Wernicke's area EP for peak N1 measured against baseline. WOOD et al. [1971] found significant left hemisphere linguistic discrimination EP differences for the negative peak, leads T3 and C3 beginning at about 105 msec and up to 200 msec [cf. WOOD, this volume]. RATLIFF and GREENBERG [1972] found the latency of a negative peak at 105–140 msec varied most between linguistic and nonlinguistic stimuli, although T5–T6 differences were not found. DORMAN [1974] found AEP correlates of speech sound discrimination for N1–P1. NEVILLE [1974] found the highest 'F' ratio for the latencies of N100 testing the verbal/nonverbal hemisphere interaction. MATSUMIYA et al. [1972], using Wernicke's leads, found their greatest EP amplitude in a negative 'W wave' at 100 msec and argue that this W wave occurs when less than 25% of the stimulus word has been heard so that it cannot be a reflection of the meaning intrinsic in the message, but rather of the mental set of the subject. Further, left hemisphere W waves increased when subjects discriminated mechanical noises, also suggesting attentional set rather than verbal processing as the source of asymmetry. The presence of some similarities between attentional and 'linguistic' EP results quoted above does not necessarily imply that only attentional factors are involved, but may suggest the possibility that attentional factors contribute significantly to the EP language laterality data in these studies.

The finding that the Wernicke (W) area leads showed the greatest selective attention effect in our study is of interest since this area, the temporo-preoccipital cortex, receives heavy cortical projections from primary visual areas [MISHKIN, 1961] and is important in discriminative learning [MILNER, 1971]. Using implanted electrodes in monkeys, CARRAN [1973] found an increased 100-msec negative EP component with a discrimination task in the infero-temporal cortex. Unilateral temporal lobectomy can greatly reduce verbal/nonverbal EP differences [FEDIO and BUCHSBAUM, 1971]. Taken together with other evidence on the neural mechanisms of attention [SKINNER and YINGLING, 1977; THOMPSON and BETTINGER, 1970], our data are consistent with a prominent role of the temporal cortex in attention processes.

Our finding of greatest attentional effects at the low intensity along with the greatest lateral asymmetry as well as direct-indirect differences further implicates attentional mechanisms. Greater attentional effects with low than with high stimulus intensities have been reported by several authors [SCHWENT et al., 1976; KOPELL et al., 1969; CHAPMAN and BRAGDON, 1964; SCHECHTER and BUCHSBAUM, 1973]. Since EP amplitude can increase with arousal as well as with attention, a kind of left hemisphere arousal of the kind suggested by KINSBOURNE [1970] could explain our results. However, the MA task, which is also a left hemisphere task, failed to increase but actually reduced the EPs. A right hemisphere distraction task would be needed to fully test the KINSBOURNE hypothesis.

The CNV as a measure of expectancy can also show larger left hemisphere responses if attentional processes are lateralized. BUTLER and GLASS [1974a] showed larger left than right central CNV not only for numeric stimuli in an arithmetic task, but for a repetitive 'checking' task less clearly a left lateralized function [cf. LOW, this volume].

Since attentional factors have not been a primary concern of most laterality research, issues much investigated in attentional research such as the effect of a variable warning period preceding the stimulus and randomization/nonrandomization of verbal/spatial stimuli have not been systematically studied. Another potentially even more serious omission in this area is a consideration of the speed/accuracy tradeoff [GREEN and LUCE, 1973]. In the reaction time or EP experiments, an individual can choose to work either for speed or accuracy and a forced-pace can overload the processing capacity [cf. DEBECKER and DESMEDT, 1970, 1971]. Instructions to the subject should be more systematically varied and made explicit. Working for speed might put a premium on attentional factors.

Summary

The influence of visual field and hemisphere of recording on visual EPs was investigated under different conditions of selective attention. 29 right-handed female subjects observed a sequence of light flashes presented to their left or right hemifields while they performed various attentional tasks or did mental arithmetic. Visual fixation was monitored to insure accurate hemifield presentation. The N140 component showed the greatest amplitude increase with attention, and this was most marked for right field presentation and left temperoparietal recording. Parallels between these results and studies of EPs obtained with verbal stimuli are reviewed. This overlap between attentional processes and verbal information processing may reflect the operation of a common attentional mechanism invoked by naive subjects confronted with logical/analytic experimental paradigms. It should be emphasized that if we examine our data from the distraction (mental arithmetic) task alone, no effect of hemispheric asymmetry or retinal site is statistically significant. In experiments where no task is assigned, subjects invent tasks which may influence hemispheric asymmetry in unpredictable ways.

Dr. M.S. BUCHSBAUM, Building 10, Room 2N212, National Institute of Mental Health, *Bethesda, MD 20014* (USA). Tel. (301) 496 4749.

Compound List of References

ADRIAN, E.D. and MATTHEWS, B.H.C.: The Berger rhythm: potential changes from the occipital lobe in man. Brain *57:* 355–385 (1934).

AGNEW, H.W.; PARKER, J.C.; WEBB, W.B., and WILLIAMS, R.L.: Amplitude measurement of the sleep electroencephalogram. Electroenceph. clin. Neurophysiol. *22:* 84–86 (1967).

AIRD, R.B. and BOWDITCH, S.: Cortical localization by electroencephalography. J. Neurosurg. *1:* 245–261 (1946).

AIRD, R.B. and ZEALEAR, D.S.: The localizing value of asymmetrical electroencephalographic tracings obtained simultaneously by homologous recording. Electroenceph. clin. Neurophysiol. *3:* 487–495 (1951).

ALAJOUANINE, T.: Aphasia and artistic realization. Brain *71:* 230–241 (1948).

ANDERSEN, P. and ANDERSSON, S.A.: Physiological basis of the alpha rhythm (Appleton Century Crofts, New York 1968).

ANDREASSI, J.L.; DE SIMONE, J.J.; FRIEND, M.A., and GROTA, P.A.: Hemispheric amplitude asymmetries in the auditory evoked potential with monaural and binaural stimulation. Physiol. Psychol. *3:* 169–171 (1975a).

ANDREASSI, J.L.; OKAMURA, H., and STERN, M.: Hemispheric asymmetries in the visual cortical evoked potential as a function of stimulus location. Psychophysiology *12:* 541–546 (1975b).

ANNETT, M.: The binomial distribution of right, mixed and left handedness. Q. Jl exp. Psychol. *19:* 327–333 (1967).

ANNETT, M.: The distribution of manual asymmetry. Br. J. Psychol. *63:* 343–358 (1972).

ARDEN, G.B.; BODIS-WOLLNER, I.; HALLIDAY, A.M.; JEFFREYS, A.; KULIKOWSKI, J.J.; SPEKREIJSE, H., and REGAN, D.: Methodology of patterned visual stimulation. Report of a working group; in DESMEDT Visual evoked potentials in man: new developments, pp. 3–15 (Oxford University Press, Oxford 1977).

ASCHOFF, J.C. and KORNHUBER, H.H.: Functional interpretation of somatic afferents in cerebellum, basal ganglia, and motor cortex; in KORNHUBER The somatosensory system, pp. 145–157 (Thieme, Stuttgart 1975).

ASLANOV, A.S.: Correlation between cortical potentials in patients with obsessive neuroses; in RUSINOV Electrophysiology of the central nervous system, pp. 39–47 (Plenum Publishing, New York, 1970).

BALI, L.; CALLAWAY, E., and NAGHDI, S.: Hemispheric asymmetry in normals and dyslexics (submitted for publication, 1977).

BALTHASAR, K.: Über das anatomische Substrat der generalisierten Tic-Krankheit (maladie des tics, Gilles de la Tourette). Arch. Psychiat. NervKrankh. *195:* 531–549 (1957).

BASSER, L.S.: Hemiplegia of early onset and the faculty of speech with special reference to the effects of hemispherectomy. Brain *85:* 427–460 (1962).

BECK, E.C.; DUSTMAN, R.E., and LEWIS, E.G.: The use of the averaged evoked potential in the evaluation of central nervous system disorders. Int. J. Neurol. *9:* 212–232 (1975).

BECKER, W.; HOEHNE, O.; IWASE, K. und KORNHUBER, H.H.: Bereitschaftspotential, prämotorische Positivierung und andere Hirnpotentiale bei sakkadischen Augenbewegungen. Vision Res. *12:* 431–436 (1972).

BEGLEITER, H.; GROSS, M.M., and KISSIN, B.: Evoked cortical responses to affective visual stimuli. Psychophysiology *3:* 336–344 (1967).

BEGLEITER, H.; GROSS, M.M.; PORJESZ, B., and KISSIN, B.: The effects of awareness on cortical evoked potentials to conditioned affective stimuli. Psychophysiology *5:* 517–529 (1969).

BEGLEITER, H. and PLATZ, H.: Cortical evoked potentials to semantic stimuli. Psychophysiology *6:* 91–100 (1969).

BENTON, A.L.: Clinical asymptomatology in right and left hemisphere lesions; in MOUNTCASTLE Interhemispheric relations and cerebral dominance, pp. 177–195 (Johns Hopkins Press, Baltimore 1962).

BERGER, H.: Über das Elektroenkephalogram des Menschen; 2. Mitteilung. J. Psychol. Neurol. *40:* 160–179 (1930).

BERGER, H.: Das Elektroenkephalogram des Menschen. Nova Acta leopoldina *6:* 173–309 (1938).

BERLUCCHI, G.; HERON, W.; HYMAN, R.; RIZZOLATTI, G., and UMILTA, C.: Simple reaction times of ipsilateral and contralateral hand to lateralized visual stimuli. Brain *94:* 419–430 (1971).

BEVER, T.G.; KIRK, R., and LACKNER, T.: An autonomic reflection of syntactic structure. Neuropsychologia *7:* 23–28 (1969).

BLUMENTHAL, A.L.: Language and psychology (Wiley, New York 1970).

BOBROW, D. and FRASER, B.: An augmented state transition network analysis procedure; in WALKER and NORTON Int. Joint Conf. on Artificial Intelligence, Washington 1969.

BODIS-WOLLNER, I.; HENLEY, C.D., and ATKIN, A.: Evaluation by evoked potentials of dissociated visual functions in patients with cerebral lesions; in DESMEDT Visual evoked potentials in man: new developments, pp. 514–524 (Oxford University Press, Oxford 1977).

BOGEN, J.E.: The other side of the brain. II. An appositional mind. Bull. Los Angeles Neurol. Soc. *34:* 135–162 (1969).

BOGEN, J.E. and Bogen, G.M.: The other side of the brain. III. The corpus callosum and creativity. Bull. Los Angeles Neurol. Soc. *34:* 191–220 (1969).

BONHOEFFER, R.: Die akuten und chronischen choreatischen Erkrankungen und die Myoklonien (Karger, Berlin 1936).

BRANCH, C.; MILNER, B., and RASMUSSEN, T.: Intracarotid sodium amytal for the lateralization of cerebral speech dominance: observations in 123 patients. J. Neurosurg. *21:* 399–405 (1964).

BRANSFORD, J.D.; BARCLAY, J., and FRANKS, J.J.: Sentence memory: constructive vs. interpretive approach. Cognit. Psychol. *3:* 193–209 (1972).

BRANSFORD, J.D. and FRANKS, J.J.: The abstraction of linguistic ideas. Cognit. Psychol. *2:* 331–350 (1971).

BRINKMAN, J. and KUYPERS, H.G.J.M.: Splitbrain monkeys: cerebral control of ipsilateral and contralateral arm, hand and finger movements. Science *176:* 536–539 (1972).

BRIX, R. and BURIAN, K.: The objectivication of verbal discrimination ability: further investigations. Audiology *12:* 481–487 (1973).

BROADBENT, D.E. and GREGORY, M.: Stimulus set and response set: the alternation of attention. Q. Jl exp. Psychol. *16:* 309–317 (1964).

BROCA, P.: Remarques sur le siège de la faculté du langage articulé suivi d'une observation d'aphémie. Bull. Soc. Anat. *6:* 330–357 (1861).

BROOKS, L.R.: An extension of the conflict between visualization and reading. Q. Jl exp. Psychol. *22:* 91–96 (1970).

BROWN, R.W.: Words and things (Free Press, Glencoe 1958).

BROWN, R.W.: Psycholinguistics (Free Press, New York 1970).

BROWN, R.W.: A first language (Harvard University Press, Cambridge 1973).

BROWN, R.W.; FRASER, C., and BELLUGI, U.: Explorations in grammar evaluation; in BELLUGI and BROWN The acquisition of language. Monogr. Soc. Res. Child Develop. *29:* 79–92 (1964).

BROWN, W.S.; MARSH, J.T., and SMITH, J.C.: Contextual meaning effects on speech-evoked potentials. Behav. Biol. *9:* 755–761 (1973).

BROWN, W.S.; MARSH, J.T., and SMITH, J.C.: Evoked potential waveform differences produced by the perception of different meanings of an ambiguous phrase. Electroenceph. clin. Neurophysiol. *41:* 113–123 (1976).

BRYDEN, M.P.: Tachistoscopic recognition, handedness and cerebral dominance. Neuropsychologia *3:* 1–8 (1965).

BUCHSBAUM, M.S.; BRYAN, J.S.; PFEFFERBAUM, A.; CARMAN, G.M., and STILLMAN, R.C.: Data acquisition system for precision eye fixation studies. Behav. Res. Meth. Instr. *5:* 411–417 (1973).

BUCHSBAUM, M. and FEDIO, P.: Visual information and evoked responses from the left and right hemispheres. Electroenceph. clin. Neurophysiol. *26:* 266–272 (1969).

BUCHSBAUM, M. and FEDIO, P.: Hemispheric differences in evoked potentials to verbal and nonverbal stimuli in the left and right visual fields. Physiol. Behav. *5:* 207–210 (1970).

BUCHWALD, J.S. and HUANG, C.M.: Far-field acoustic response: origins in the cat. Science *189:* 382–384 (1975).

BUTLER, R.A.; KEIDEL, W.D., and SPRENG, M.: An investigation of the human cortical evoked potential under conditions of monaural and binaural stimulation. Acta oto-lar. *68:* 317–326 (1969).

BUTLER, S.A. and GLASS, A.: Interhemispheric asymmetry of contingent negative variation during numeric operation. Electroenceph. clin. Neurophysiol. *30:* 366 (1971).

BUTLER, S.R. and GLASS, A.: Asymmetries in the CNV over left and right hemispheres while subjects await numeric information. Biol. Psychol. *2:* 1–16 (1974a).

BUTLER, S.R. and GLASS, A.: Asymmetries in the electroencephalogram associated with cerebral dominance. Electroenceph. clin. Neurophysiol. *36:* 481–491 (1974b).

CALLAWAY, E.: Correlations between averaged evoked potentials and measures of intelligence. Archs gen. Psychiat. *29:* 553–558 (1973).

CALLAWAY, E. and HALLIDAY, R.A.: Evoked potential variability: effects of age, amplitude and methods of measurement. Electroenceph. clin. Neurophysiol. *34:* 125–133 (1973).

CALLAWAY, E. and HARRIS, P.R.: Coupling between cortical potentials from different areas. Science *183:* 873–875 (1974).

CALLAWAY, E. and STONE, G.C.: Evoked response for the study of intelligence. Agressologie *10:* 262–281 (1969).

CALMES, R.L. and CRACCO, R.Q.: Comparison of somatosensory and somatomotor evoked responses to median nerve and digital nerve stimulation. Electroenceph. clin. Neurophysiol. *31:* 547–562 (1971).

CARMELIET, J.; DEBECKER, J., and DESMEDT, J.E.: A random interval generator using beta ray emission. Electroenceph. clin. Neurophysiol. *30:* 354–356 (1971).

CARRAN, A.B.: Inferotemporal evoked potentials after visual discrimination training in squirrel monkey. Percept. Mot. Skills *36:* 851–854 (1973).

CASE, T.J. and BUCY, P.C.: Localization of cerebral lesions by electroencephalography. J. Neurophysiol. *1:* 245–261 (1938).

CAZDEN, C.B.: Environmental assistance to the child's acquisition of grammar; unpubl. diss. Harvard University (1965).

CHAPMAN, R.M.: Psychological variables in AEP experiment; in DONCHIN and LINDSLEY Average evoked potentials. NASA SP-191, pp. 262–275 (US Government Printing Office, Washington 1969).

CHAPMAN, R.M.: Evoked potentials of the brain related to thinking; in McGUIGAN and SCHOONOVER The psychophysiology of thinking, pp. 69–108 (Academic Press, New York 1973).

CHAPMAN, R.N. and BRAGDON, H.R.: Evoked responses to numerical and non-numerical visual stimuli while problem-solving. Nature, Lond. *203:* 1155–1157 (1964).

CHARNIAK, E.: Toward a model of children's story comprehension, AI TR-266 (MIT Artificial Intelligence Laboratory, Cambridge 1972).

CHARTOK, H.E.; GLASSMAN, P.R.; POON, L.W., and MARSH, G.R.: Changes in alpha rhythm asymmetry during learning of verbal and visuospatial tasks. Physiol. Behav. *15:* 237–239 (1975).

CHASE, R.A.: Neurological aspects of language disorders in children; in IRWIN and MARGE Principles of childhood language disabilities, pp. 99–135 (Appleton, New York 1972).

CHI, J.G.; DOOLING, E.C., and GILLES, F.H.: Gyral development of the human brain. Annls Neurol. *1:* 86–93 (1977).

CHOMSKY, N.: Syntactic structures (Mouton, The Hague 1957).

CHOMSKY, N.: Aspects of the theory of syntax (MIT Press, Cambridge 1965).

CHOMSKY, N.: Cartesian linguistics (Harper & Row, New York 1966).

CHOMSKY, N.: Language and mind (Harcourt, Brace & World, New York 1968).

CHOMSKY, N.: Language and mind (Harcourt Brace Jovanovich, New York 1972).

COHEN, J.: Very slow brain potentials relating to expectancy: the CNV; in DONCHIN and LINDSLEY Average evoked potentials. NASA SP-191, pp. 143–198 (US Government Printing Office, Washington 1969).

COHEN, J.: CNV and visual recognition. Electroenceph. clin. Neurophysiol. *33:* suppl., pp. 201–204 (1973).

COHEN, H.D.; ROSEN, R.C., and GOLDSTEIN, L.: Electroencephalographic laterality changes during human sexual orgasm. Archs sex. Behav. *5:* 189–199 (1976).

COHEN, J. and WALTER, W.G.: The interaction of responses in the brain to semantic stimuli. Psychophysiology *2:* 187–196 (1966).

COHN, R.: Differential cerebral processing of noise and verbal stimuli. Science *172:* 599–601 (1971).

CONEL, J.L.: The postnatal development of the human cerebral cortex: the cortex of the one-month infant (Harvard University Press, Cambridge 1941).

CONEL, J.L.: The postnatal development of the human cerebral cortex: the cortex of the six-year child (Harvard University Press, Cambridge 1967).

COOPER, F.S.: Spectrum analysis. J. acoust. Soc. Am. *22:* 761–762 (1950).

COOPER, R.; OSSELTON, J.W., and SHAW, J.C.: EEG Technology; 2nd ed. (Butterworth, London 1974).

COOPER, W.E.: Selective adaption to speech; in RESTLE, SHIFFRIN, CASTELLAN, LINDMAN and PISONI Cognitive theory, vol. 1 (Erlbaum, Potomac 1975).

CORNIL, L. et GASTAUT, H.: Etude électroencéphalographique de la dominance sensorielle d'un hémisphère cérébral. Presse méd. *37:* 421–422 (1947).

CRACCO, R.Q.: The initial positive potential of the human scalp-recorded somatosensory evoked response. Electroenceph. clin. Neurophysiol. *32:* 623–629 (1972).

CRAIK, F.I.M. and LOCKHART, R.S.: Levels of processing: a framework for memory research. J. verbal learning Behav. *11:* 671–684 (1972).

CROWELL, D.H.; JONES, R.H.; KAPUNIAI, L.E., and NAKAGAWA, J.K.: Unilateral cortical activity in newborn humans: an early index of cerebral dominance? Science *180:* 205–208 (1973).

CROWELL, D.H.; KAPUNIAI, L.E., and GARBANATI, J.A.: Hemispheric differences in the rhythmic responses to photic stimulation in human infants; in DESMEDT Visual evoked potentials in man: new developments, pp. 353–362 (Oxford University Press, Oxford 1977).

CULVER, C.M.; TANLEY, J.C., and EASON, R.G.: Evoked cortical potentials: relation to hand dominance and eye dominance. Percept. Mot. Skills *30:* 407–414 (1970).

CUTTING, J.E.: Two left-hemisphere mechanisms in speech perception. Percept. Psychophys. *16:* 601–612 (1974).

DAVIDSON, R.J. and SCHWARTZ, G.E.: Patterns of lateralization during cardiac biofeedback versus the self-regulation of emotion: sex differences. Psychophysiology *13:* 62–68 (1976).

DAVIS, A.E.: Power spectral analysis of flash and click evoked responses. Electroenceph. clin. Neurophysiol. *35:* 287–291 (1973).

DAVIS, A.E. and CRAPPER, D.R.: The spatio-temporal basis of hemispheric asymmetries. Brain Lang. (submitted, 1977).

DAVIS, A.E. and WADA, J.A.: Hemispheric asymmetry: frequency analysis of visual and auditory evoked responses to nonverbal stimuli. Electroenceph. clin. Neurophysiol. *37:* 1–9 (1974).

DAVIS, A.E. and WADA, J.A.: Hemispheric asymmetries in human infants. Brain Lang. *4:* 23–31 (1977a).

Davis, A.E. and Wada, J.A.: Lateralization of speech dominance by spectral analysis of evoked potentials. J. Neurol. Neurosurg. Psychiat. 40: 1–4 (1977b).

Davis, A.E. and Wada, J.A.: Speech dominance and handedness in the normal human. Brain Lang. (accepted, 1977c).

Davis, A.E. and Wada, J.A.: Hemispheric asymmetries of visual and auditory information processing. Neuropsychologia (accepted, 1977d).

Davis, H.: Brain stem and other responses in electric response audiometry. Ann. Otol. Rhinol. Lar. 85: 1–12 (1976).

Day, R.S. and Bartlett, J.C.: Separate speech and nonspeech processing in dichotic listening? J. acoust. Soc. Am. 51: 79 (1972).

Day, R.S. and Wood, C.C.: Interactions between linguistic and non-linguistic processing. J. acoust. Soc. Am. 51: 79 (1972).

Debecker, J. and Desmedt, J.E.: Rate of intermodality switching disclosed by sensory evoked potentials averaged during signal detection tasks. J. Physiol., Lond. 185: 52–53 P (1966).

Debecker, J. and Desmedt, J.E.: Maximum capacity for sequential one-bit auditory decisions. J. exp. Psychol. 83: 366–372 (1970).

Debecker, J. and Desmedt, J.E.: Cerebral evoked potential correlates in forced-paced tasks. Nature new Biol. 234: 118–120 (1971).

Deecke, L.; Becker, W.; Grözinger, B.; Scheid, P., and Kornhuber, H.H.: Human brain potentials preceding voluntary limb movements. Electroenceph. clin. Neurophysiol. Suppl. 33: 87–94 (1973).

Deecke, L.; Englitz, H.G.; Kornhuber, H.H., and Schmitt, G.: Cerebral potentials preceding voluntary movement in patients with bilateral or unilateral Parkinson akinesia; in Desmedt Attention, voluntary contraction and event-related cerebral potentials, pp. 151–163 (Karger, Basel 1977).

Deecke, L.; Grözinger, B., and Kornhuber, H.H.: Voluntary finger movements in man: cerebral potentials and theory. Biol. Cybern 23: 99–119 (1976).

Deecke, L. and Kornhuber, H.H.: Cerebral potentials and the initiation of voluntary movement; in Desmedt Attention, voluntary contraction and event-related cerebral potentials, pp. 132–150 (Karger, Basel 1977).

Deecke, L.; Scheid, P., and Kornhuber, H.H.: Distribution of readiness potential, pre-motion positivity, and motor potential of the human cerebral cortex preceding voluntary finger movements. Expl Brain Res. 7: 158–168 (1969).

D'Elia, G. and Perris, C.: Cerebral function dominance and depression. An analysis of EEG amplitude in depressed patients. Acta psychiat. scand. 49: 191–197 (1973).

Denhyer, K. and Barrett, B.: Selective loss of visual information in STM by means of visual and verbal interpolated tasks. Psychol. Sci. 25: 100–102 (1971).

DeRenzi, E.; Faglioni, P., and Scotti, G.: Hemispheric contribution to exploration of space through the visual and tactile modality. Cortex 6: 191–203 (1970).

Desmedt, J.E.: Somatosensory cerebral evoked potentials in man; in Rémond Handbook of electroencephalography and clinical neurophysiology, vol. 9, pp. 55–82 (Elsevier, Amsterdam 1971).

Desmedt, J.E.: Physiological studies of the efferent recurrent auditory system; in Keidel and Neff Handbook of sensory physiology, vol. 5, part 2, pp. 219–246 (Springer, Berlin 1975).

DESMEDT, J.E. (ed.): Visual evoked potentials in man: new developments (Oxford University Press, Oxford 1977a).

DESMEDT, J.E. (ed.): Attention, voluntary contraction and event-related cerebral potentials (Karger, Basel 1977b).

DESMEDT, J.E.: Some observations on the methodology of cerebral evoked potentials in man; in DESMEDT Attention, voluntary contraction and event-related cerebral potentials, pp. 12–29 (Karger, Basel 1977c).

DESMEDT, J.E.: Active touch exploration of extrapersonal space elicits specific electrogenesis in the right cerebral hemisphere of intact right-handed man. Proc. natn. Acad. Sci. USA 74: 4037–4040 (1977d).

DESMEDT, J.E.; BRUNKO, E., and DEBECKER, J.: Maturation of the somatosensory evoked potentials in normal infants and children, with particular reference to the early N_1 component. Electroenceph. clin. Neurophysiol. 40: 43–58 (1976).

DESMEDT, J.E.; BRUNKO, E.; DEBECKER, J., and CARMELIET, J.: The system bandpass required to avoid distortion of early components when averaging somatosensory evoked potentials. Electroenceph. clin. Neurophysiol. 37: 407–410 (1974).

DESMEDT, J.E. and DEBECKER, J.: Cerebral event-related potentials in man: dissociation of the cognitive P400 component from slow negative shifts. Brain Res. (1978).

DESMEDT, J.E.; DEBECKER, J. et MANIL, J.: Mise en évidence d'un signe électrique cérébral associé à la détection par le sujet d'un stimulus sensoriel tactile. Bull. Acad. R. Méd. Belg. 5: 887–936 (1965).

DESMEDT, J.E. and ROBERTSON, D.: Differential enhancement of early and late components of the cerebral somatosensory evoked potentials during fast sequential cognitive tasks in man. J. Physiol., Lond. 271: 761–782 (1977).

DESMEDT, J.E.; ROBERTSON, D.; BRUNKO, E., and DEBECKER, J.: Somatosensory decision tasks in man: early and late components of the cerebral potentials evoked by stimulation of different fingers in random sequences. Electroenceph. clin. Neurophysiol. 43: 403–414 (1977).

DEVILLIERS, J.: Quantitative aspects of agrammatism in aphasia. Cortex 10: 36–54 (1974).

DIDE, M.: Les désorientations temroro-spatiales et la prépondérance de l'hémisphère droit dans les anoso-akinésies proprioceptives. Encéphale 33: 276–294 (1938).

DIMOND, S.J. and BEAUMONT, J.G. (eds): Hemisphere function in the human brain (Elek, London 1974).

DIXON, W.J. (ed.): BMD: biomedical computer programs (University of California Press, Berkeley 1973).

DONALD, M.W.: Limits on current theories of transient evoked potentials; in DESMEDT Cognitive components in cerebral event-related potentials and selective attention. Prog. clin. Neurophysiol., vol. 6 (Karger, Basel, in press 1978).

DONCHIN, E.: A multivariate approach to the analysis of average evoked potentials. IEEE Trans. biomed. Enging 13: 131–139 (1966).

DONCHIN, E.: Data analysis techniques in evoked potential research; in DONCHIN and LINDSLEY Average evoked potentials. NASA SP-191, pp. 199–236 (US Government Printing Office, Washington 1969).

DONCHIN, E.: Brain electrical correlates of pattern recognition; in INBAR Signal analysis and pattern recognition in biomedical engineering, pp. 199–218 (Wiley, New York 1975).

DONCHIN, E.; CALLAWAY, E.; COOPER, R.; DESMEDT, J.E.; GOFF, W.R.; HILLYARD, S.A., and SUTTON, S.: Publication criteria for studies of evoked potentials (EP) in man. Report of a committee; in DESMEDT Attention, voluntary contraction and event-related cerebral potentials, pp. 1–11 (Karger, Basel 1977a).

DONCHIN, E. and COHEN, L.: Averaged evoked potentials and intramodality selective attention. Electroenceph. clin. Neurophysiol. 22: 537–546 (1967).

DONCHIN, E.; GERBRANDT, L.; LEIFER, L., and TUCKER, L.: Is the contingent negative variation contingent on a motor response? Psychophysiology 9: 178–188 (1972).

DONCHIN, E.; JOHNSON, R.; HERNONG, R., and KUTAS, M.: Covariation of the magnitude of the CNV and P300 as a function of the subject's task; in McCALLUM and KNOTT The responsive brain, pp. 76–80 (Wright, Bristol 1976).

DONCHIN, E.; KUBOVY, M.; KUTAS, M.; JOHNSON, R., jr., and HERNING, R.I.: Graded changes in evoked response (P300) amplitude as a function of cognitive activity. Percept. Psychophys. 14: 319–324 (1973).

DONCHIN, E.; KUTAS, M., and McCARTHY, G.: Comparison of the hemispheric asymmetries of the readiness potential and CNV. Proc. Psychonomic Soc. 15th Annu. Meet., Boston 1974.

DONCHIN, E.; KUTAS, M., and McCARTHY, G.: Electrocortical indices of hemispheric utilization; in HARNAD, DOTY, GOLDSTEIN, JAYNES and KRAUTHAMER Lateralization in the nervous system, pp. 339–384 (Academic Press, New York 1977b).

DONCHIN, E. and LINDSLEY, D.B. (eds): Averaged evoked potentials. NASA SP-191 (US Government Printing Office, Washington 1969).

DONCHIN, E.; TUETING, P.; RITTER, W.; KUTAS, M., and HEFFLEY, E.: On the independence of the CNV and the P300 components of the averaged evoked potential. Electroenceph. clin. Neurophysiol. 38: 449–461 (1975).

DORMAN, M.F.: Auditory evoked potential correlates of speech and sound discrimination. Percept. Psychophys. 15: 215–220 (1974).

DOYLE, J.C.; ORNSTEIN, R., and GALIN, D.: Lateral specialization of cognitive mode: EEG frequency analysis. Psychophysiology 11: 567–578 (1974).

DROHOCKI, Z.: L'intégrateur de l'électroproduction cérébrale pour l'électroencéphalographie quantitative. Revue neurol. 80: 619–624 (1948).

DUMAS, R. and MORGAN, A.: EEG asymmetry as a function of occupation, task, and task difficulty. Neuropsychologia 13: 219–228 (1975).

DUMENKO, V.N.: Electroencephalographic investigation of cortical responses in dogs during formation of a conditional reflex stereotype; in RUSINOV Electrophysiology of the central nervous system, pp. 107–127 (Plenum Publishing, New York, 1970).

DUNCAN-JOHNSON, C. and DONCHIN, E.: On quantifying surprise: the variation of event-related potentials with subjective probability. Proc. Soc. for Psychophysiological Res., San Diego 1976.

EASON, R.G.; GROVES, P.; WHITE, C.T., and ODEN, D.: Evoked cortical potentials: relation to visual field and handedness. Science 156: 1643–1646 (1967).

EIMAS, P.D.: Speech perception in early infancy; in COHEN and SALAPATEK Infant perception from sensation to cognition, vol. 2, pp. 193–231 (Academic Press New York 1975).

EIMAS, P.D. and CORBIT, J.D.: Selective adaptation of linguistic feature detectors. Cognit. Psychol. 4: 99–109 (1973).

EIMAS, P.D.; SIQUELAND, E.R.; JUSCZYK, P., and VIGORITO, J.: Speech perception in infants. Science *171:* 303–306 (1971).

ERTL, J.P. and SCHAFER, E.W.P.: Cortical activity preceding speech. Life Sci. *6:* 473–479 (1967).

ERTL, J. and SCHAFER, E.W.P.: Cortical activity preceding speech. Life Sci. *8:* 539 (1969).

ETTLINGER, G.; WARRINGTON, E., and ZANGWILL, O.L.: A further study of visual-spatial agnosia. Brain *80:* 335–361 (1957).

FEDIO, P. and BUCHSBAUM, M.: Unilateral temporal lobectomy and changes in evoked responses during recognition of verbal and nonverbal material in the left and right visual fields. Neuropsychologia *9:* 261–271 (1971).

FELDMAN, R.M. and GOLDSTEIN, R.: Averaged evoked responses to synthetic syntax sentences. J. Speech Hear. Res. *10:* 689–696 (1967).

FILBEY, R.A. and GAZZANIGA, M.S.: Splitting the normal brain with reaction time. Psychon. Sci. *17:* 335 (1969).

FILLENBAUM, S.: Psycholinguistics. Annu. Rev. Psychol. *22:* 251–308 (1971).

FLOR-HENRY, P.: Lateralized temporal-limbic dysfunction and psychopathology. Ann. N.Y. Acad. Sci. *280:* 777–795 (1976).

FOA, U.G. and SCHLESINGER, I.M.: Syntactical errors in sentence recall. A re-analysis of Mehler's data. Unpubl. paper, Israel Inst. App. Social Res. (1965).

FODOR, J.A.; BEVER, T.G., and GARRETT, M.F.: The psychology of language (McGraw-Hill, New York 1974).

FODOR, J.A. and GARRETT, M.: Competence and performance; in LYONS and WALES Psycholinguistics papers (Edinburgh Press, Edinburgh 1966).

FODOR, J.A. and GARRETT, M.: Competence and performance. Behav. Biol. *9:* 755–761 (1973).

FONTENOT, D.J. and BENTON, A.L.: Tactile perception of direction in relation to hemispheric locus of lesion. Neuropsychologia *9:* 83–88 (1971).

FREDERIKS, J.A.M.: Disorders of the body schema; in VINKEN and BRUYN Handbook of clinical neurology, vol. 4, pp. 207–240 (North Holland, Amsterdam 1969).

FRIEDMAN, D.; SIMSON, R.; RITTER, W., and RAPIN, I.: Cortical evoked potentials elicited by real speech words and human sounds. Electroenceph. clin. Neurophysiol. *38:* 13–19 (1975a).

FRIEDMAN, D., SIMSON, R.; RITTER, W., and RAPIN, I.: The late positive component (P300) and information processing in sentences. Electroenceph. clin. Neurophysiol. *38:* 255–262 (1975b).

FRITZELL, B.: The velopharyngeal muscles in speech: an electromyographic and cineradiographic study. Acta oto-lar. Suppl. *250:* 81 (1969).

FROMKIN, V.; KRASHEN, S.; CURTISS, S.; RIGLER, D., and RIGLER, M.: The development of language; in GENIE A case of language acquisition beyond the critical period. Brain Lang. *1:* 81–107 (1974).

GAINOTTI, G.: Emotional behavior and hemispheric side of the lesion. Cortex *8:* 41–55 (1972).

GALAMBOS, R.; BENSON, P.; SMITH, T.S.; SHULMAN-GALAMBOS, C., and OSIER, H.: On hemispheric differences in evoked potentials to speech stimuli. Electroenceph. clin. Neurophysiol. *39:* 279–283 (1975).

GALAMBOS, R. and HECOX, K.: Clinical applications of the brain stem auditory evoked potentials; in DESMEDT Auditory evoked potentials in man. Psychopharmacology correlates of EPs, pp. 1–19 (Karger, Basel 1977).

GALIN, D. and ELLIS, R.R.: Asymmetry in evoked potentials as an index of lateralized cognitive processes: relation to EEG alpha asymmetry. Psychophysiology 13: 45–50 (1975).

GALIN, D. and ORNSTEIN, R.: Lateral specialization of cognitive mode: an EEG study. Psychophysiology 9: 412–418 (1972).

GALIN, D. and ORNSTEIN, R.: Individual differences in cognitive style. I. Reflective eye movements. Neuropsychologia 12: 367–376 (1974).

GARDNER, B.T. and GARDNER, R.A.: Two-way communication with an infant chimpanzee; in SCHRIER and STOLLNITZ Behavior of nonhuman primates, vol. 4, pp. 117–184 (Academic Press, New York 1971).

GARNER, W.R.: The processing of information and structure (Erlbaum, Potomac 1974).

GAVRILOVA, N.A.: Spatial synchronization of cortical potentials in patients with disturbances of association; in RUSINOV Electrophysiology of the central nervous system, pp. 129–143 (Plenum Publishing, New York 1970).

GAZZANIGA, M.S.: The bisected brain (Appleton Century Crofts, New York 1970).

GAZZANIGA, M.S.: Cerebral dominance viewed as a decision system; in DIMOND and BEAUMONT Hemisphere function in the human brain, pp. 367–382 (Elek, London 1974).

GAZZANIGA, M.S.; BOGEN, J.E., and SPERRY, R.W.: Observations on visual perception after disconnection of the cerebral hemispheres in man. Brain 88: 221–236 (1965).

GAZZANIGA, M.S. and HILLYARD, S.A.: Language and speech capacity of the right hemisphere. Neuropsychologia 9: 273–280 (1971).

GAZZANIGA, M.S. and HILLYARD, S.A.: Attention mechanisms following brain bisection; in KORNBLUM Attention and performance, vol. 4, pp. 221–238 (Academic Press, New York 1973).

GERBRANDT, L.K.; GOFF, W.R., and SMITH, D.B.: Distribution of the human average movement potential. Electroenceph. clin. Neurophysiol. 34: 461–474 (1973).

GESCHWIND, N.: Disconnexion syndromes in animals and man. Brain 88: 237–294 and 585–644 (1965).

GESCHWIND, N.: The organization of language and the brain. Science 170: 940–944 (1970).

GESCHWIND, N.: Selected papers on language and the brain (Reidel, Dordrecht 1974).

GESCHWIND, N. and LEVITSKY, W.: Human brain: Left-right asymmetries in temporal speech region. Science 161: 186–187 (1968).

GEVINS, A.S.; YEAGER, C.L.; DIAMOND, S.L.; SPIRE, J.P.; ZEITLIN, G.M., and GEVINS, A.H.: Automated analysis of the electrical activity of the human brain (EEG). A progress report. Proc. IEEE 63: 1382–1399 (1975).

GIANNITRAPANI, D.: Developing concepts of lateralization of cerebral functions. Cortex 3: 353–370 (1967).

GIANNITRAPANI, D. and DARROW, C.W.: Differences in EEG time relationships in right and left handed individuals. Electroenceph. clin. Neurophysiol. 15: 721P (1963).

GIANNITRAPANI, D.; DARROW, C.W., and SORKIN, A.: Asleep and awake interhemispheric EEG phase relationships in left- and right-handed subjects. Am. Psychol. 19: 480–481 (1964).

GIANNITRAPANI, D.; SORKIN, A.I., and ENNENSTEIN, J.: Laterality preference of children and adults as related to interhemispheric EEG phase activity. J. neurol. Sci. *3:* 139–150 (1966).

GIBBS, F.A. and GIBBS, E.L.: Atlas of Electroencephalography (Addison-Wesley, Reading, Mass. 1952).

GIBBS, F.A.; LENNOX, W.G., and GIBBS, E.L.: The electro-encephalogram in diagnosis and in localization of epileptic seizures. Archs Neurol. Psychiat., Chicago *36:* 1225–1235 (1936).

GIBLIN, D.R.: Somatosensory evoked potentials in healthy subjects and in patients with lesions of the nervous system. Ann. N. Y. Acad. Sci. *112:* 93–142 (1964).

GIBSON, J.J.: Observations on active touch. Psychol. Rev. *69:* 477–491 (1962).

GILDEN, L.; VAUGHAN, H.C., and COSTA, L.D.: Summated human EEG potentials associated with voluntary movement. Electroenceph. clin. Neurophysiol. *20:* 433–438 (1966).

GIPS, J.; PFEFFERBAUM, A., and BUCHSBAUM, M.: Use of a small process control computer in a psychophysiological laboratory. Psychophysiology *8:* 538–542 (1971).

GLANVILLE, A.D. and ANTONITIS, J.J.: The relationship between occipital alpha activity and laterality. J. exp. Psychol. *49:* 294–299 (1955).

GOFF, W.R.; MATSUMIYA, Y.; ALLISON, T., and GOFF, G.D.: Cross-modality comparisons of averaged evoked potentials; in DONCHIN and LINDSLEY Average evoked potentials. NASA SP-191, pp. 95–118 (US Government Printing Office, Washington 1969).

GOFF, W.R.; ROSNER, B.S., and ALLISON, T.: Distribution of cerebral somatosensory evoked responses in normal man. Electroenceph. clin. Neurophysiol. *14:* 697–713 (1962).

GOLDSTEIN, L.; MURPHREE, H.B.; SUGERMAN, A.A.; PFEIFER, C.C., and JENNEY, E.H.: Quantitative electroencephalographic analysis of naturally occurring (Schizophrenic) and drug-induced psychotic states in human males. Clin. Pharmac. Ther. *4:* 10–21 (1963).

GOLDSTEIN, L. and STOLTZFUS, N.W.: Psychoactive drug-induced changes of EEG amplitude relationships. Agents Actions *3:* 124–132 (1973).

GOLDSTEIN, L.; STOLZFUS, N.W., and GARDOCKI, J.F.: Changes in interhemispheric amplitude relationships in the EEG during sleep. Physiol. Behav. *8:* 811–815 (1972).

GOLDSTEIN, L.; SUGERMAN, A.A.; MARJERISON, G., and STOLTZFUS, N.W.: The EEG and differential hemispheric functions. Bucke Memorial Soc., Montreal 1973.

GOLDSTEIN, R.: Electroencephalographic audiometry; in JERGER Modern development in audiology, pp. 407–435 (Academic Press, New York 1973).

GOLLA, F.; HUTTON, E.L., and WALTER, W.G.: The objective study of mental imagery. Physiological concomitants. J. ment. Sci. *89:* 216–223 (1943).

GOODGLASS, H.: Discussion paper on cortical functioning; in GILBERT Speech and cortical functioning (Academic Press, New York 1972).

GOODGLASS, H. and BLUMSTEIN, S.: Psycholinguistics and aphasia (Johns Hopkins Press, Baltimore 1973).

GOTT, P.S. and BOYARSKY, L.L.: The relationship of cerebral dominance and handedness to visual evoked potentials. J. Neurobiol. *3:* 65–77 (1972).

GOUGH, P.: Grammatical transformations and speed of understanding. J. verbal Learn. Behav. *4:* 107–111 (1965).

GRABOW, J.D. and ELLIOTT, W.: The role of the averaged glossokinetic response in the electrophysiologic assessment of hemispheric asymmetries during speech. Electroenceph. clin. Neurophysiol. *34:* 758–759 (1973).

GRABOW, J.D. and ELLIOTT, F.W.: The electrophysiologic assessment of hemispheric asymmetries during speech. J. Speech. Hear. Res. *17:* 64–72 (1974).

GREEN, D.M. and LUCE, R.D.: Speed-accuracy trade off in auditory detection; in KORNBLUM Attention and performance, vol. 4 (Academic Press, New York 1973).

GREENBERG, J. (ed.): Universals of language (MIT Press, Cambridge 1962).

GREENBERG, H.J. and GRAHAM, J.T.: Electroencephalographic changes during learning of speech and non-speech stimuli. J. verbal Learn. Behav. *9:* 274–281 (1970).

GREENBERG, H.J. and METTING, P.J.: Averaged encephalic response of aphasics to linguistic and non-linguistic auditory stimuli. J. Speech Hear. Res. *17:* 113–124 (1974).

GRÖZINGER, B.; KORNHUBER, H.H., and KRIEBEL, J.: Inter- and intra-hemispheric asymmetries of brain potentials preceding speech and phonation. Electroenceph. clin. Neurophysiol. *34:* 737–738 (1973).

GRÖZINGER, B.; KORNHUBER, H.H., and KRIEBEL, J.: Respiration correlated brain potentials. J. interdiscipl. Cycle Res. *5:* 287–294 (1974a).

GRÖZINGER, B.; KORNHUBER, H.H., and KRIEBEL, J.: Methodological problems in the investigation of cerebral potentials preceding speech: determining the onset and suppressing artefacts caused by speech. Neuropsychologia *13:* 263–270 (1975).

GRÖZINGER, B.; KORNHUBER, H.H., and KRIEBEL, J.: EEG investigation of hemispheric asymmetries preceding speech. The R-wave; in McCALLUM and KNOTT The responsive brain, pp. 103–107 (Wright, Bristol 1976).

GRÖZINGER, B.; KORNHUBER, H.; KRIEBEL, J. und MURATA, K.: Menschliche Hirnpotentiale vor dem Sprechen. Pflügers Arch. ges. Physiol., *332:* suppl., p. R 100, 200 (1972).

GRÖZINGER, B.; KORNHUBER, H.H.; KRIEBEL, J., and MURATA, K.: Cerebral potentials during respiration and preceding vocalization. Electroenceph. clin. Neurophysiol. *36:* 435 (1974b).

GUR, R.E.; GUR, R.C., and HARRIS, L.J.: Cerebral activation as measured by subjects lateral eye movements is influenced by experimenter location. Neuropsychologia *13:* 35–44 (1975).

HAALAND, K.Y.: The effect of dichotic, monaural and diotic verbal stimuli on auditory evoked potentials. Neuropsychologia *12:* 339–345 (1974).

HALLIDAY, A.M.: Changes in the form of cerebral evoked responses in man associated with various lesions of the nervous system. Electroenceph. clin. Neurophysiol. Suppl. *25:* 178–192 (1967).

HANNAN, E.J.: Multiple time series analysis (Wiley, New York 1970).

HARMONY, J.; RICARDO, G.; OTERO, FERNANDEZ, S., and VALDES, P.: Symmetry of the visual evoked response in normal human subjects. Electroenceph. clin. Neurophysiol. *35:* 237–240 (1973).

HARNAD, S.R.: Interhemispheric division of labor. R.M. Bucke Memorial Soc., Montreal 1973.

HARNAD, S.R.; DOTY, R.W.; GOLDSTEIN, L.; JAYNES, J., and KRAUTHAMER, G. (eds.): Lateralization in the nervous system, pp. 537 (Academic Press, New York 1977).

HARNAD, S. R.; STEKLIS, H. D., and LANCASTER, J. (eds.): Origins and evolution of language and speech. Ann. N.Y. Acad. Sci. *280:* 1–914 (1976).

HARRIS, C. M.: A study of the building blocks in speech. J. acoust. Soc. Am. *25:* 962–969 (1953).

HASSLER, R. and DIECKMANN, G.: Relief of obsessive-compulsive disorders, phobias and tics by stereotactic coagulation of the rostral intralaminar and medial-thalamic nuclei; in LAITINEN and LIVINGSTON Surgical approaches in psychiatry, pp. 206–212 (Medicine & Technical Publishing, Lancaster 1973).

HAZEMANN, P.; OLIVIER, L. et DUPONT, E.: Potentiels évoqués somesthésiques recueillis sur le scalp chez 6 hémisphérectomisés. Revue neurol. *12:* 246–257 (1969).

HEAD, H.: Aphasia and kindred disorders of speech (Cambridge University Press, Cambridge 1926).

HÉCAEN, H.: Clinical symptomatology in right and left hemispheric lesions; in MOUNT-CASTLE Interhemispheric relations and cerebral dominance, pp. 215–243 (Johns Hopkins Press, Baltimore 1962).

HÉCAEN, H. and AJURIAGUERRA, J.: Left handedness, manual superiority and cerebral dominance (Grune & Stratton, New York 1964).

HÉCAEN, H.; PENFIELD, W.; BERTRAND, C., and MALMO, R.: The syndrome of apractognosia due to lesions of the minor cerebral hemisphere. Archs Neurol. Psychiat., Chicago *75:* 400–434 (1956).

HÉCAEN, H. and SAUGUET, J.: Cerebral dominance in left-handed subjects. Cortex *7:* 19–48 (1971).

HEESCHEN, C. and JURGENS, R.: Pragmatic-semantic and syntactic factors influencing ear differences in dichotic listening. Cortex *13* (1977).

HEWITT, W.: The development of the corpus callosum. J. Anat. *96:* 355–358 (1962).

HILLYARD, S. A.: The CNV and human behavior: a review. Electroenceph. clin. Neurophysiol., *33:* suppl., pp. 162–172 (1973).

HILLYARD, S. A.; HINK, R. F.; SCHWENT, V. L., and PICTON, T. W.: Electrical signs of selective attention in the human brain. Science *182:* 177–180 (1973).

HILLYARD, S. A. and PICTON, T. W.: Event-related brain potentials and selective information processing in man; in DESMEDT Cognitive components in cerebral event-related potentials and selective attention (Karger, Basel, in press 1978).

HILLYARD, S. A.; SQUIRES, K. C.; BAUER, J. W., and LINDSAY, P. H.: Evoked potential correlates of auditory signal detection. Science *172:* 1357–1360 (1971).

HOLMES, V. and FORSTER, K.: Detection of extraneous signals during sentence recognition. Percept. Psychophys. *7:* 297–301 (1970).

HOOVEY, Z. B.; HEINEMANN, U., and CREUTZFELD, O. D.: Inter-hemispheric synchrony of alpha waves. Electroenceph. clin. Neurophysiol. *32:* 337–347 (1972).

HUMPHREY, M. E.: Consistency of hand usage: a preliminary enquiry. Br. J. Educat. Psychol. *21:* 214–224 (1951).

INGVAR, D.: Studies of cerebral blood flow during speech and reading. Electroenceph. clin. Neurophysiol. *36:* 561–576 (1974).

INGVAR, D. H.: Brain activation patterns revealed by measurements of regional cerebral blood flow; in DESMEDT Cognitive components in cerebral event-related potentials and selective attention. Prog. clin. Neurophysiol., vol. 6 (Karger, Basel, in press 1978).

JAMES, W.E.; MEFFERD, R.B., and WIELAND, B.A.: Repetitive psychometric measures: handedness and performance. Percept. Mot. Skills 25: 209–212 (1967).

JARVILEHTO, T. and FRUHSTORFER, H.: Differentiation between slow cortical potentials associated with motor and mental acts in man. Expl. Brain Res. 11: 309–317 (1970).

JASPER, H.H.: The ten twenty electrode system. Electroenceph. clin. Neurophysiol. 10: 371–375 (1958).

JENKINS, G.M. and WATTS, D.G.: Spectral analysis and its applications (Holden-Day, San Francisco 1968).

JEWETT, D.; ROMANO, H.N., and WILLISTON, J.S.: Human auditory evoked responses: possible brain stem components detected on the scalp. Science 167: 1517–1518 (1970).

JOHNSON, R.E. and DONCHIN, E.: Equivocation, feedback and the amplitude of P300. Proc. Soc. for Psychophysiological Res., San Diego 1976.

JONES, E.G. and POWELL, T.P.S.: Anatomical organization of the somatosensory cortex; in IGGO Somatosensory system. Handbook of sensory physiology, vol. 2, pp. 579–620 (Springer, Berlin 1973).

KAPLAN, R.: A general syntactic processor; in RUSTIN Natural language processing (Prentice-Hall, Englewood Cliffs 1973).

KARLIN, L.: Cognition, preparation, and sensory-evoked potentials. Psychol. Bull. 73: 122-136 (1970).

KATZ, J.J.: The philosophy of language (Harper & Row, New York 1966).

KEATING, L.W. and RUHM, H.B.: Some observations on the effects of attention to stimuli on the amplitude of the acoustically evoked response. Audiology 10: 177–184 (1971).

KELLOGG, W.N.: Communication and language in the home-raised chimpanzee. Science 162: 423–427 (1968).

KENT, R.D.; Carney, P.J., and Severeid, L.R.: Velar movement and timing: evaluation of a model for binary control. J. Speech Hear. Res. 17: 470–488 (1974).

KIMURA, D.: Cerebral dominance and the perception of verbal stimuli. Can. J. Psychol. 15: 166–171 (1961).

KIMURA, D.: Left-right differences in the perception of melodies. Q. Jl exp. Psychol. 16: 355–358 (1964).

KIMURA, D.: Dual functional asymmetry of the brain in visual perception. Neuropsychologia 4: 275–285 (1966).

KINSBOURNE, M.: The cerebral basis of lateral asymmetries in attention. Acta psychol. 33: 193–201 (1970).

KINSBOURNE, M.: The minor cerebral hemisphere as a source of aphasic speech. Archs Neurol., Chicago 25: 302–306 (1971).

KINSBOURNE, M.: Eye-head turning indicates cerebral lateralization. Science 176: 539–541 (1972).

KINSBOURNE, M.: The control of attention by interaction between the cerebral hemispheres; in KORNBLUM Attention and performance, vol. 4, pp. 239–256 (Academic Press, New York 1973).

KINSBOURNE, M.: Direction of gaze and distribution of cerebral thought processes. Neuropsychologia 12: 279–281 (1974).

KINSBOURNE, M. and SMITH, W.L. (eds): Hemispheric disconnection and cerebral function (Thomas, Springfield 1974).

KLASS, D. and BICKFORD, R.G.: Glossokinetic potentials appearing in the electroencephalogram. Electroenceph. clin. Neurophysiol. *12:* 239 (1960).

KOCEL, K.; GALIN, D.; ORNSTEIN, R., and MERRIN, E.: Lateral eye movement and cognitive mode. Psychon. Sci. *27:* 223–224 (1972).

KOENIG, W.; DUNN, H.K., and LACY, L.Y.: The sound spectrograph. J. acoust. Soc. Am. *17:* 19–49 (1946).

KOOI, K.A.; ECKMAN, H.G., and THOMAS, M.H.: Observations on the response to photic stimulation in organic cerebral dysfunction. Electroenceph. clin. Neurophysiol. *9:* 239–250 (1957).

KOOI, K.A.; Guvener, A.M., and Bagchi, B.K.: Visual evoked responses in lesions of the optic pathway. Electroenceph. clin. Neurophysiol. *18:* 524 (1965).

KOPELL, B.S.; WITTNER, W.K., and WARRICK, G.L.: The effects of stimulus differences light intensity and selective attention on the amplitude of the visual average evoked potential in man. Electroenceph. clin. Neurophysiol. *26:* 619–622 (1969).

KORNHUBER, H.H.: Motor functions of cerebellum and basal ganglia: the cerebellocortical saccadic (ballistic) clock, the cerebello-nuclear hold regulator and the basal ganglia ramp (voluntary speed smooth movement) generator. Kybernetik *8:* 157–167 1971).

KORNHUBER, H.H.: Neural control of input into long term memory: limbic system and amnestic syndrome in man; in ZIPPEL Memory and transfer of information, pp. 1–22 (Plenum Publishing, New York 1973).

KORNHUBER, H.H.: Cerebral cortex, cerebellum and basal ganglia: an introduction to their motor functions; in SCHMITT and WORDEN The neurosciences third study program, pp. 267–280 (MIT Press, Cambridge 1974a).

KORNHUBER, H.H.: The vestibular system and the general motor system; in KORNHUBER Vestibular system. Handbook of sensory physiology, vol. 6, 2, pp. 581–620 (Springer, Berlin 1974b).

KORNHUBER, H.H. und DEECKE, L.: Hirnpotentialänderungen beim Menschen vor und nach Willkürbewegungen, dargestellt mit Magnetbandspeicherung und Rückwärtsanalyse. Pflügers Arch. ges. Physiol. *281:* 52 (1964).

KORNHUBER, H.H. und DEECKE, L.: Hirnpotentialänderungen bei Willkürbewegungen und passiven Bewegungen des Menschen: Bereitschaftspotential und reafferente Potentiale. Pflügers Arch. ges. Physiol. *284:* 1–17 (1965).

KRAYENBUEHL, H.; AKERT, K.; HARTMANN-V. MONAKOW, K. et YASARGIL, M.G.: Etude de la corrélation anatomoclinique chez des malades opérés de parkinsonisme. Neurocirurgia, Santiago *10:* 397–412 (1964).

KRYNICKI, V.E.: Hemispheric differences in form recognition. A behavioral and evoked potential study; Columbia University, New York (1976).

KURTZBERG, D. and VAUGHAN, H.G.: Electrocortical potentials associated with eye movements; in ZITZMUND The oculomotor system and brain function, pp. 135–145 (Slovak Academy of Sciences, Bratislava 1973).

KURTZBERG, D. and VAUGHAN, H.G.: Electrophysiological observations on the visuomotor system and visual neurosensorium; in DESMEDT Visual evoked potentials in man: new developments, pp. 314–331 (Oxford University Press, Oxford 1977).

KUTAS M. and DONCHIN, E.: Studies of squeezing: handedness, responding hand, response force, and asymmetry of readiness potential. Science *186:* 545–548 (1974).

KUTAS, M. and DONCHIN, E.: The effect of handedness, of responding hand, and of response force on the contralateral dominance of the readiness potential; in DESMEDT Attention, voluntary contraction and event-related cerebral potentials, pp. 189–210 (Karger, Basel 1977).

KUTAS, M.; McCARTHY, G., and DONCHIN, E.: Augmenting mental chronometry: the P300 as a measure of stimulus evaluation time. Science *197:* 792–795 (1977).

KUYPERS, H.G. J.M.: The anatomical organization of the descending pathways and their contributions to motor control especially in primates; in DESMEDT New developments in electromyography and clinical neurophysiology, vol. 3, pp. 38–68 (KARGER, Basel 1973).

LAIRY, G.C.; RÉMOND, A.; RIEGER, H., and LESÈVRE, N.: The alpha average. Clinical application in children. Electroenceph. clin. Neurophysiol. *26:* 453–467 (1969).

LANSDELL, H.: Verbal and non-verbal factors in right hemisphere speech. J. comp. Psychol. *69:* 738 (1969).

LANSING, R.W. and BARLOW, J.S.: Rhythmic after-activity to flashes in relation to the background alpha which precedes and follows the photic stimuli. Electroenceph. clin. Neurophysiol. *32:* 149–160 (1972).

LANSING, R.W. and THOMAS, H.: The laterality of photic driving in normal adults. Electroenceph. clin. Neurophysiol. *16:* 290–294 (1964).

LAPPIN, J.S. and DISCH, K.: The latency operating characteristics: effects of stimulus probability on choice reaction time. J. exp. Psychol. *92:* 419–427 (1972).

LASHLEY, K.S.: The problem of serial order in behavior; in BEACH, HEBB, MORGAN and NISSEN The neuropsychology of Lashley, pp. 506–528 (McGraw-Hill, New York 1960).

LEE, P.R.; ORNSTEIN, R.E.; GALIN, D.; DEIKMAN, A., and TART, C.T.: Symposium on Consciousness, pp. 53–66 (Viking Press, New York 1976).

LEISSNER, P.; LINDHOLM, L.E., and PETERSEN, I.: Alpha amplitude dependence on skull thickness as measured by ultra sound technique. Electroenceph. clin. Neurophysiol. *32:* 149–160 (1970).

LENNEBERG, E.H.: Understanding language without ability to speak. A case report. J. abnorm. Psychol. *65:* 419–425 (1962).

LENNEBERG, E.H.: Biological foundation of language (Wiley, New York 1967).

LENNEBERG, E.H.: The neurology of language. Deadalus *102:* 115–134 (1973).

LEVY, J.: Possible basis for the evolution of lateral specialization of the human brain. Nature, Lond. *224:* 614–615 (1969).

LEVY, J.; NEBES, R., and SPERRY, R.W.: Expressive language in the surgically separated minor hemisphere. Cortex *7:* 49–58 (1971).

LHERMITTE, F. and GAUTIER, J.C.: Aphasia; in VINKEN and BRUYN Handbook of clinical neurology, vol. 4, pp. 84–104 (North-Holland, Amsterdam 1969).

LIBERMAN, A.M.: The grammar of speech and language. Cognit. Psychol. *1:* 301–323 (1970) (1970).

LIBERMAN, A. M.; COOPER, F.S.; SHANKWEILER, D., and STUDDERT-KENNEDY, M.: Perception of the speech code. Psychol. Rev. *74:* 431–461 (1967).

LIBERMAN, A.M.; DELATTRE, P.C., and COOPER, F.S.: The role of selected stimulus variables in the perception of the unvoiced-stop consonants. Am. J. Psychol. *65:* 497–516 (1952).

LIEBERMAN, P.: On the origins of language. An introduction to the evolution of human speech (Macmillan, New York 1975).

LINDSLEY, D.B.: Bilateral differences in brain potentials from the two cerebral hemispheres in relation to laterality and stuttering. J. exp. Psychol. *26:* 211–225 (1940).

LISKE, E.; HUGHES, H.M., and STOWE, D.E.: Cross-correlation of human alpha activity: normative data. Electroenceph. clin. Neurophysiol. *22:* 429–436 (1966).

LISKER, L. and ABRAMSON, A.S.: A cross-language study of voicing in initial stops: acoustical measurements. Word *20:* 384–422 (1964).

LIVANOV, M.N.; GAVRILOVA, N.A., and ASLANOV, A.S.: Intercorrelations between different cortical regions of human brain during mental activity. Neuropsychologia *2:* 281–289 (1964).

LOVELESS, N.E. and SANFORD, A.J.: Slow potentials correlates of preparatory sets. Biol. Psychol. *1:* 303–314 (1974).

LOVELESS, N.E. and SANFORD, A.J.: The impact of warning signal intensity on reaction time and components of the contingent negative variation. Biol. Psychol. *2:* 217–226 (1975).

LOW, M.D.: Very slow potentials relating to expectancy, the CNV-discussion; in DONCHIN and LINDSLEY Average evoked potentials. NASA SP–191, pp. 163–177 (US Government Printing Office, Washington 1969).

LOW, M.D.; BORDA, R.P.; FROST, J.D., and KELLAWAY, P.: Surface-negative slow-potential shift associated with conditioning in man. Neurology *16:* 771–782 (1966).

LOW, M.D. and STOILEN, L.: CNV and EEG in children: maturational characteristics and findings in the MCD syndrome. Electroenceph. clin. Neurophysiol. *33:* suppl., pp. 139–143 (1973).

LOW, M.; WADA, J.A., and FOX, M.: Electroencephalographic localization of the conative aspects of language production in the human brain. Trans. Am. neurol. Ass. *98:* 129–133 (1973).

LOW, M.D.; WADA, J.A., and FOX, M.: Electroencephalographic localization of the cognitive aspects of language production in the human brain; in McCALLUM and KNOTT The responsive brain, pp. 165–168 (Wright, Bristol 1976).

LURIA, A.R.: Higher cortical functions in man (Basic Books, New York 1966).

LURIA, A.R.: Language and brain: Brain Lang. *1:* 1–14 (1974).

LURIA, A.R. and YODOVICH, F.I.A.: Speech and the development of mental processes in the child (Staples, London 1959).

MACKAY, D.M. (ed.): Evoked brain potentials as indicators of sensory information processing. Neurosci. Res. Prog. Bull. *7:* 181–276 (1969).

MACKWORTH, N.H.: A stand camera for line-of-sight recording. Percept. Psychophys. *2:* 119–127 (1967).

MACLEAN, P.D.: Sensory and perceptive factors in emotional functions of the triune brain; in GRENELL and GABAY Biological foundations of psychiatry, pp. 177–198 (Raven Press, New York 1976).

MACNAMARA, J.: Cognitive basis of language learning in infants. Psychol. Rev. *79:* 1–13 (1972).

MACNEILAGE, P.: Motor control of serial ordering of speech. Psychol. Rev. *77:* 182–196 (1970).

MAJKOWSKI, J.; BOCHENEK, Z.; BOCHENEK, W.; KNAPIK-FIJALKOWSKA, D., and KOPEC, J.:

Latency of averaged evoked potentials to contralateral and ipsilateral auditory stimulation in normal subjects. Brain Res. *25:* 416–419 (1971).

MARKS, L.E.: Some structural and sequential factors in the processing of sentences. Verbal Learn. Behav. *6:* 707–713 (1967).

MARSH, G.R. and Thompson, L.W.: Hemispheric asymmetry in slow cortical potentials as a function of verbal or non-verbal set. Soc. Neurosci. Abstr., p. 170 (1972).

MARSH, G.R. and THOMPSON, L.W.: Effect of verbal and non-verbal psychological set on hemispheric asymmetries in the CNV. Electroenceph. clin. Neurophysiol. *33:* suppl., pp. 195–200 (1973).

MARSH, J.T.; BROWN, W.S., and SMITH, J.C.: Far-field frequency-following responses: correlates of low pitch auditory perception in humans. Electroenceph. clin. Neurophysiol. *38:* 113–119 (1975).

MARSH, J.T. and WORDEN, F.G.: Sound evoked frequency-following responses in the central auditory pathway. Laryngoscope *78:* 1149–1163 (1968).

MATSUMIYA, Y.; TAGILIASCO, V.; LOMBROSO, C.T., and GOODGLASS, H.: Auditory evoked response: meaningfulness of stimuli and interhemispheric asymmetry. Science *175:* 790–792 (1972).

MATTHEWS, W.A.: Transformational complexity and short term recall. Language Speech *11:* 120–128 (1968).

MATTINGLY, I.G.; Liberman, A.M.; SYDRAL, A.K., and HALWES, T.: Discrimination in speech and non-speech modes. Cognit. Psychol. *2:* 131–157 (1971).

McADAM, D.W. and SEALES, D.M.: Bereitschaftspotential enhancement with increased level of motivation. Electroenceph. clin. Neurophysiol. *27:* 73–75 (1969).

McADAM, D.W. and WHITAKER, H.A.: Language production: electroencephalographic localization in the normal human brain. Science *172:* 499–502 (1971).

McCALLUM, W.C. and KNOTT, J.R. (eds): Event related slow potentials of the brain. Electroenceph. clin. Neurophysiol. *33:* suppl., pp. 390 (1973).

McCALLUM, W.C. and KNOTT, J.R. (eds): The responsive brain (Wright, Bristol 1976).

McCALLUM, W.C. and PAPAKOSTOPOULOS, D.: Slow potential changes in human brainstem associated with preparation for decision or action; in Abstr. Int. Symp. on Cerebral Evoked Potentials in Man, Brussels 1974.

McCARTHY, G. and DONCHIN, E.: Dissociating the components of the CNV in a visual matching task. Soc. Psychophysiol. Res., San Diego 1976.

McKEE, G.; HUMPHREY, B., and McADAM, D.W.: Scaled lateralization of alpha activity during linguistic and musical tasks. Psychophysiology *10:* 441–443 (1973).

McKEEVER, W.F. and HULING, M.: Left cerebral hemisphere superiority in tachistoscopic word recognition performance. Percept. Mot. Skills *30:* 763–766 (1970).

McMAHON, L.: Grammatical analysis as part of understanding a sentence; unpubl. diss. Harvard University (1963).

McNEILL, D.: On theories of language acquisition; in DIXON and HORTON Verbal behavior and general behavior theory, pp. 406–420 (Prentice-Hall, Englewood Cliffs 1968).

McNEILL, D.: The acquisition of language (Harper & Row, New York 1970).

MEHLER, J.: Some effects of grammatical transformations on the recall of English sentences. Verbal Learn. Behav. *2:* 346–351 (1963).

MILLER, E.: Handedness and the pattern of human ability. Br. J. Psychol. *62:* 111–112 (1971).

MILLER, G.A.: The magical number seven plus or minus two. Psychol. Rev. *63:* 81–97 (1956).

MILLER, G.A.: Some preliminaries to psycholinguistics. Am. Psychol. *20:* 15–20 (1965).

MILLER, G.A. and McKEAN, K.: A chronometric study of some relations between sentences. Q. Jl exp. Psychol. *16:* 297–308 (1964).

MILLER, G.A. and McNEILL, G.D.: Psycholinguistics; in LINDZEY and ARONSON Handbook of social psychology, vol. 3 (Addison-Wesley, Reading, Mass. 1968).

MILNER, B.: Interhemispheric differences in the localization of psychological processes in man. Br. med. Bull. *27:* 272–277 (1971).

MILNER, B.: Hemispheric specialization: scope and limits; in SCHMITT and WORDEN The neurosciences Third Study Programm, pp. 75–89 (MIT Press, Cambridge 1974).

MILNER, B. and TAYLOR, L.: Right hemisphere superiority in tactile pattern recognition after cerebral commissurotomy: evidence for non-verbal memory. Neuropsychologia *10:* 1–15 (1972).

MIRON, M.S. and Osgood, C.E.: Language behavior: the multivariate structure of qualification; in CATTELL Handbook of multivariate experimental psychology, pp. 790–819 (Rand-McNally, Chicago 1966).

MISHKIN, M.: Perceptual learning considered as a function of posterior association cortex. Acta psychol. *19:* 337–338 (1961).

MISHKIN, M.: Possible link between interhemispheric integration in monkeys and cerebral dominance in man; in MOUNTCASTLE Interhemispheric relations and cerebral dominance, pp. 123–126 (John Hopkins Press, Baltimore 1962).

MOFFITT, A.R.: Consonant cue perception by twenty to twenty-four week old infants. Child Dev. *42:* 717–731 (1971).

MOLFESE, D.L.: Cerebral asymmetry in infants, children and adults: auditory evoked responses to speech and music stimuli; diss., Pennsylvania State, University Park (1972).

MOLFESE, D.L.: Some acoustic cues responsible for differential hemispheric responding in infants. Expl Child Psychol. *22* (1976).

MOLFESE, D.L.; FREEMAN, R.B., and PALERMO, D.S.: The ontogeny of brain lateralization for speech and nonspeech stimuli. Brain Lang. *2:* 356–368 (1975).

MORGAN, A.H.; MAC DONALD, H., and HILGARD, E.R.: EEG alpha: lateral asymmetry related to task and hypnotizability. Psychophysiology *11:* 275–282 (1974).

MORGAN, A.H.; McDONALD, P.J., and McDONALD, H.: Differences in bilateral alpha activity as a function of experimental task with a note on lateral eye movements and hypnotizability. Neuropsychologia *9:* 459–469 (1971).

MORRELL, L.K. and HUNTINGTON, D.A.: Electrocortical cortical localization of language production. Science *174:* 1359–1360 (1971).

MORRELL, L.K. and HUNTINGTON, D.A.: Cortical potentials time-locked to speech production: evidence for probable cerebral origin. Life Sci. *11:* 921–929 (1972).

MORRELL, L.K. and SALAMY, J.G.: Hemispheric asymmetry of electrocortical response to speech stimuli. Science *174:* 164–166 (1971).

MOUNTCASTLE, V.B. (ed.): Interhemispheric relations and cerebral dominance (John Hopkins Press, Baltimore 1962).

MOUNTCASTLE, V.B.; LYNCH, J.C.; GEORGOPOULOS, A.; SAKATA, H., and ACUNA, C.: Posterior parietal association cortex of the monkey: command functions for operations within extrapersonal space. J. Neurophysiol. *38:* 871–908 (1975).

MOUSHEGIAN, G.: The frequency-following potentials in man; in DESMEDT Auditory evoked potentials in man. Psychopharmacology correlates of EPs, pp. 20–29 (Karger, Basel 1977).

MOUSHEGIAN, G.; RUPERT, A.L., and STILLMAN, R.D.: Scalp-recorded early responses in man to frequencies in the speech range. Electroenceph. clin. Neurophysiol. *35:* 665–667 (1973).

NEBES, R.D.: Superiority of the minor hemisphere in commissurotomized man for the perception of part-whole relations. Cortex *7:* 333–349 (1971).

NEBES, R.D.: Dominance of the minor hemisphere in commissurotomized man on a test of figural unification. Brain *95:* 633–638 (1972).

NEBES, R.D.: Dominance of the minor hemisphere in commissurotomized man for the perception of part-whole relationships; in KINSBOURNE and SMITH Hemispheric disconnection and cerebral function, pp. 155–164 (Thomas, Springfield 1974).

NELSON, J.M.; PHILLIPS, R., and GOLDSTEIN, L.: Interhemispheric EEG laterality relationships following psychoactive agents and during operant performance in rabbits; in HARNAD, DOTY, GOLDSTEIN, JAYNES and KRAUTHAMER Lateralization in the nervous system, pp. 451–470 (Academic Press, New York 1977).

NEVILLE, H.: Electrographic correlates of lateral asymmetry in the processing of verbal and nonverbal auditory stimuli. J. Psychol. Res. *3:* 151–163 (1974).

NEWCOMBE, F.: Missile wounds of the brain: a study of psychological deficits (Oxford University Press, Oxford 1969).

NICKERSON, R.S.: Binary-classification reaction time: a review of some studies of human information-processing capabilities. Psychon. Monogr. Suppl. *4:* whole No. 65 (1971).

NIELSEN, J.M.: Agnosia, apraxia, aphasia: their value in cerebral localization (Hoeber, New York 1946).

NORMAN, D.A. and BOBROW, D.G.: On data-limited and resource-limited processes. Cognit. Psychol. *7:* 44–64 (1975).

OHLRICH, E.S. and BARNET, A.B.: Auditory evoked responses during the first year of life. Electroenceph. clin. Neurophysiol. *32:* 161–169 (1972).

OLDFIELD, R.C.: The assessment and analysis of handedness: the Edinburgh inventory. Neuropsychologia *9:* 97–113 (1971).

ORNITZ, E.M.; RITVO, E.R.; PANMAN, L.E.; LEE, Y.H.; CARR, E.M., and WALTER, R.D.: The auditory evoked response in normal and autistic children during sleep. Electroenceph. clin. Neurophysiol. *25:* 221–230 (1965).

ORNITZ, E.M.; RITVO, E.R.; LEE, Y.H.; PANMAN, L.M.; WALTER, R.D., and MASON, A.: The auditory evoked response in babies during REM sleep. Electroenceph. clin. Neurophysiol. *27:* 195–198 (1967).

ORNSTEIN, R.E. and GALIN, D.: Physiological studies of consciousness (Institute for Cultural Research, Tunbridge Wells 1973).

ORNSTEIN, R.E. and GALIN, D.: Physiological studies of consciousness; in LEE, ORNSTEIN, GALIN, DEIKMAN and TART Symposium on consciousness, pp. 53–66 (Viking Press, New York 1976).

OSGOOD, C.E.: The nature and measurement of meaning. Psychol. Bull. *49:* 197–237 (1952).

OSGOOD, C.E.: Semantic differential technique in the comparative study of cultures. Am. Anthrop. *66:* 171–200 (1964).

OSGOOD, C.E.: Exploration in semantic space: a personal diary. J. soc. Issues *27:* 5–64 (1971).

OSGOOD, C.E.; SUCI, G.J., and TANNENBAUM, P.: The measurement of meaning (University of Illinois Press, Urbana 1957).

OTTO, D.A. and LEIFER, L.J.: The effect of modifying response and performance on the CNV in humans. Electroenceph. clin. Neurophysiol. *33:* suppl., pp. 29–37 (1973).

PACHELLA, R.G.: The interpretation of reaction time in information processing research; in KANTOWITZ Human information processing: tutorials in performance and cognition (Erlbaum, Potomac 1974).

PAIVIO, A.; YUILLE, J.C., and MADIGAN, S.A.: Concreteness, imagery and meaningfulness values for 925 nouns. J. exp. Psychol. *76:* 1–25 (1968).

PANDYA, D.N. and VIGNOLO, L.A.: Interhemispheric projections of the parietal lobe in the rhesus monkey. Brain Res. *15:* 49–65 (1969).

PAPCUN, G.; KRASHEN, S.; TERBEEK, D.; REMINGTON, R., and HARSHMAN, R.: Is the left hemisphere specialized for speech, language and/or something else? J. acoust. Soc. Am. *55:* 319–327 (1974).

PAUL, D. and SUTTON, S.: Evoked potential correlates of response criterion in auditory signal detection. Science *177:* 362–364 (1972).

PENFIELD, W. and RASMUSSEN, T.: The cerebral cortex of man (MacMillan, New York 1950).

PENFIELD, W. and ROBERTS, L.: Speech and brain mechanisms (Princeton University Press, Princeton 1959).

PERONNET, F. and MICHEL, F.: The asymmetry of the auditory evoked potentials in normal man and in patients with brain lesions; in DESMEDT Auditory evoked potentials in man. Psychopharmacology correlates of EPs, pp. 130–141 (Karger, Basel 1977).

PERONNET, F.; MICHEL, F.; ECHALLIER, J.F., and GIROD, J.: Coronal topography of human auditory evoked responses. Electroenceph. clin. Neurophysiol. *37:* 225–230 (1974).

PERRY, N.W. and CHILDERS, D.G.: The human visual evoked response: method and theory (Thomas, Springfield 1969).

PETERS, J.F. and MENDEL, M.I.: Early components of the averaged electroencephalographic response to monaural and binaural stimulation. Audiology *13:* 195–204 (1974).

PETERSON, G.E.; WANG, W.S.-Y., and SIVERTSEN, E.: Segmentation techniques in speech synthesis. J. acoust. Soc. Am. *30:* 739–742 (1958).

PIAGET, J.: Piaget's theory; in MUSSEN Carmichael's handbook of child psychology (Wiley, New York 1970).

PICTON, T.W.; HILLYARD, S.A.; KRAUSZ, H.I., and Galambos, R.: Human auditory evoked potentials. I. Evaluation of components. Electroenceph. clin. Neurophysiol. *36:* 179–190 (1974).

PISONI, D.B.: On the nature of categorical perception of speech sounds; unpubl. diss., Ann Arbor (1971).

PISONI, D.B.: Auditory and phonetic memory codes in the discrimination of consonants and vowels. Percept. Psychophys. *13:* 253–260 (1974).

POON, L.W.; THOMPSON, L.W.; WILLIAMS, R.B., jr., and MARSH, G.: Changes of antero-posterior distribution of CNV and late positive component as a function of information processing demands. Psychophysiology *11:* 660–673 (1974).

PREMACK, D.: Language in chimpanzee? Science *172:* 808–822 (1971).

PRICE, L.L. and GOLDSTEIN, R.: Average evoked responses for measuring auditory sensitivity in children. J. Speech Hear. Dis. *31:* 248–256 (1966).

PRICE, L.L.; ROSENBLUT, B.; GOLDSTEIN, R., and SHEPHERD, D.C.: The averaged evoked response to auditory stimulation. J. Speech Hear. Res. *9:* 361–370 (1966).

PROVINS, K.A. and CUNLIFFE, P.: The relationship between EEG activity and handedness. Cortex *8:* 136–146 (1972).

RABINER, L.R. and GOLD, B.: Theory and application of digital signal processing (Prentice-Hall, Englewood Cliffs 1975).

RANEY, E.T.: Brain potentials and lateral dominance in identical twins. J. exp. Psychol. *24:* 21–39 (1939).

RATLIFF, S.S. and GREENBERG, H.J.: The averaged encephalic response to linguistic and nonlinguistic auditory stimuli. J. audit. Res. *12:* 14–25 (1972).

REGAN, D.: Evoked potentials in psychology, sensory physiology and clinical medicine (Chapman & Hall, London 1972).

REGAN, D.: Evoked potential indications of the processing of pattern, colour and depth information; in DESMEDT Visual evoked potentials in man: new developments, pp. 234–249 (Oxford University Press, Oxford 1977).

RÉMOND, A.; LESÈVRE, N. JOSEPH, J.P.; RIEGER, H., and LAIRY, G.C.: The alpha average. I. Methodology and description. Electroenceph. clin. Neurophysiol. *26:* 245–265 (1969).

RHODES, L.E.; DUSTMAN, R.E., and BECK, E.C.: The visual evoked response: a comparison of bright and dull children. Electroenceph. clin. Neurophysiol. *27:* 364–372 (1969).

RHODES, L.E.; OBITZ, F.W., and CREEL, D.: Effect of alcohol and task on hemispheric asymmetry of visually evoked potentials in man. Electroenceph. clin. Neurophysiol. *38:* 561–568 (1975).

RICHLIN, M.; WEINSTEIN, M., and WEISINGER, M.: Development of neurophysiological indices of retardation: interhemispheric asymmetry of visual evoked cortical response. Intern. J. Neurosci. *6:* 257–261 (1976).

RICHLIN, M.; WEISINGER, M.; WEINSTEIN, S.; GIANNINI, M., and MORGENSTERN, M.: Interhemispheric asymmetries of evoked cortical responses in retarded and normal children. Cortex *7:* 98–104 (1971).

RITTER, W.; SIMSON, R., and VAUGHAN, H.G.: Association cortex potentials and reaction time in auditory discrimination. Electroenceph. clin. Neurophysiol. *33:* 547–555 (1972).

RIZZOLATTI, G.; UMILTA, C., and BERLUCCHI, G.: Opposite superiorities of the right and left cerebral hemispheres in discriminative reaction time to physiognomical and alphabetical material. Brain *94:* 431–442 (1971).

ROBBINS, K.I. and MCADAM, D.W.: Interhemispheric alpha asymmetry and imagery mode. Brain Lang. *1:* 189–193 (1974).

ROCHFORD, J.M.; SWARTBURG, M.; CHOWDHREY, S.M., and GOLDSTEIN, L.: Some quantitative EEG correlates of Psychopathology. Res. Comm. Psychol. Psychiat. Behav. *1:* 211–226 (1976).

ROHRBAUGH, J.W.; SYNDULKO, K., and LINDSLEY, D.B.: Brain wave components of the contingent negative variation in humans. Science *191:* 1055–1057 (1976).

ROSENZWEIG, M.R.: Representations of the two ears at the auditory cortex. Am. J. Physiol. *167:* 147–158 (1951).

ROSSI, G.F. and ROSADINI, G.: Experimental analysis of cerebral dominance in man; in MILLIKAN and DARLEY Brain mechanisms underlying speech and language, pp. 167–184 (Grune & Stratton, New York 1967).

ROTH, W.T.; KOPELL, B.S., and BERTOZZI, D.E.: The effect of attention on the average evoked response to speech sounds. Electroenceph. clin. Neurophysiol. *29:* 38–46 (1970).

RUBENS, A.B.; MAHOWALD, M.W., and HUTTON, J.T.: Asymmetry of the lateral (sylvian) fissures in man. Neurology *26:* 620–624 (1976).

RUBIN, D.C.: The subjective estimation of relative syllabe frequency. Percept. Psychophys. *16:* 193–196 (1974).

RUCHKIN, D.S. and SUTTON, S.: Latency characteristics and trial-by-trial variations of emitted cerebral potentials; in DESMEDT Cognitive components in cerebral event-related potentials and selective attention. Progr. clin. Neurophysiol., vol. 6 (Karger, Basel, in press 1978).

RUHM, H.B.: Lateral specificity of acoustically evoked EEG responses. Non-verbal, non-meaningful stimuli. J. audit. Res. *11:* 1–8 (1971).

RUMBAUGH, D.M.; GLASERFIELD, E. VON; WARNER, H.; PISANI, P., and GILL, T.V.: Lana (chimpanzee) learning language: a progress report. Brain Lang. *1:* 205–212 (1974).

RUOFF, P.; DOERR, H.O.; FULLER, P.W.; MARTIN, D.C., and RUOFF, L.O.: Motor and cognitive interactions in the lateral cerebral functions in children. An EEG study. Soc. psychophysiol. Res. San Diego 1976.

SAKODA, J.M.; COHEN, B.H., and BEALL, G.: Test of significance for a series of statistical tests. Psychol. Bull. *51:* 172–175 (1954).

SAMRA, K.; RIKLAN, M.; LEVITA, E.; ZIMMERMAN, J.; WALTZ, J.; BERGMANN, L., and COOPER, J.S.: Language and speech correlates of anatomically verified lesions in thalamic surgery for parkinsonism. J. Speech Hear. Res. *12:* 510–540 (1969).

SATZ, P.; ACHENBACH, K., and FENNELL, E.: Correlations between assessed manual laterality and predicted speech laterality in a normal population. Neuropsychologia *5:* 295–310 (1967).

SAVIN, H. and PERCHONOCK, E.: Grammatical structures and the immediate recall of English sentences. J. verbal Learn. Behav. *4:* 348–353 (1965).

SCHAFER, E.W.P.: Cortical activity preceding speech: semantic specificity. Nature, Lond. *216:* 1338–1339 (1967).

SCHALTENBRAND, G.: The effects on speech and language of stereotactical stimulation in thalamus and corpus callosum. Brain Lang. *2:* 70–77 (1975).

SCHECHTER, G. and BUCHSBAUM, M.: The effects of attention, stimulus intensity and individual differences on the averaged evoked response. Psychophysiology *10:* 392–400 (1973).

SCHENKENBERG, T. and DUSTMAN, R.E.: Visual, auditory, and somatosensory evoked response changes related to age, hemisphere and sex. Proc. Am. Psychol. Ass. 1970: 183–184.

SCHIMMEL, H.: The ($\pm$) reference: accuracy of estimated mean components in average response studies. Science *157:* 92–94 (1967).

SCHUELL, H.; JENKINS, J.J., and JIMENEZ-PABON, E.: Aphasia in adults (Harper & Row, New York 1964).

SCHWARTZ, G.E.; DAVIDSON, R.J., and MAER, F.: Right hemisphere lateralization for emotion in the human brain: interaction with cognition. Science *190:* 286–288 (1975).

SCHWARTZ, G.E.; DAVIDSON, R.J., and PUGASH, E.: Voluntary control of patterns of EEG parietal asymmetry: cognitive concomitants. Psychophysiology *13:* 498–504 (1976).

SCHWARTZ, G.; DAVIDSON, R.J.; MAER, F., and BROMFIELD, E.: Patterns of hemispheric dominance in musical, emotional, verbal and spatial tasks. Soc. Psychophysiol. Res., New Orleans 1973.

SCHWENT, V.L. and HILLYARD, S.A.: Evoked potential correlates of selective attention with multi-channel auditory inpunts. Electroenceph. clin. Neurophysiol. *38:* 131–138 (1975).

SCHWENT, V.L.; HILLYARD, S.A., and GALAMBOS, R.: Selective attention and the auditory vertex potential. Effects of signal intensity and masking noise. Electroenceph. clin. Neurophysiol. *40:* 615–622 (1976).

SEGALOWITZ, S.J. and GRUBER, F.A. (eds.): Language development and neurological theory (Academic Press, New York 1977).

SELNES, O.A.: The corpus callosum: some anatomical and functional consideration with special reference to language. Brain Lang. *1:* 111–139 (1974).

SEMMES, J.: A non-tactual factor in astereognosis. Neuropsychologia *3:* 295–315 (1965).

SEMMES, J.: Hemispheric specialization. A possible clue to mechanism. Neuropsychologia *6:* 11–26 (1968).

SEMMES, J.; WEINSTEIN, S.; GHENT, L., and TEUBER, H.L.: Correlates of impaired orientation in personal and extrapersonal space. Brain *86:* 747–772 (1963).

SHAGASS, C.: Evoked brain potentials in psychiatry (Plenum Publishing, New York 1972).

SHANKWEILER, D.P. and STUDDERT-KENNEDY, M.: Identification of constants and vowels presented to the left and right ears. Q. Jl exp. Psychol. *19:* 59–63 (1967).

SHANKWEILER, D. and STUDDERT-KENNEDY, M.: A continuum of lateralization for speech perception? Brain Lang. *2:* 212–225 (1975).

SHANNON, C.E. and WEAVER, W.: The mathematical theory of commun. (University of Illinois Press, Urbana 1949).

SHELBURNE, S.A.: Visual evoked responses to word and nonsense syllabe stimuli. Electroenceph. clin. Neurophysiol. *32:* 17–25 (1972).

SHELBURNE, S.A.: Visual evoked responses to language stimuli in normal children. Electroenceph. clin. Neurophysiol. *34:* 135–143 (1973).

SHUCARD, D.W.; SHUCARD, J.L., and THOMAS, D.G.: Auditory evoked potentials, habituation and cerebral lateralization of cognitive processing. Soc. Psychophysiol. Res., San Diego 1976.

SILVERMAN, A.J.; ADEVAL, G., and MCGOUGH, W.E.: Some relationships between handedness and perception. J. psychosom. Res. *10:* 151–158 (1966).

SKINNER, B.F.: Verbal behavior (Appleton Century Crofts, New York 1957).

SKINNER, J.E. and YINGLING, C.D.: Central gating mechanisms that regulate event-related potentials and behavior. A neural model for attention; in DESMEDT Attention, voluntary contraction and event-related cerebral potentials, pp. 30–70 (Karger, Basel 1977).

SMITH, A.: Speech and other functions after left (dominant) hemispherectomy. J. Neurol. Neurosurg. Psychiat. *29:* 467–471 (1966).

SMITH, J.C.; MARSH, J.T., and BROWN, W.S.: Far-field recorded frequency-following responses: evidence for the locus of brainstem sources. Electroenceph. clin. Neurophysiol. *39:* 465–472 (1975).

SNEDECOR, G.W. and COCHRAN, W.G.: Statistical methods (Iowa State University Press, Ames 1967).

SNIDER, J.G. and OSGOOD, C.E.: Semantic differential technique: a sourcebook (Aldine Atherton, Chicago 1969).

SPEKREIJSE, H.; ESTEVEZ, O., and REITS, D.: Visual evoked potentials and the physiological analysis of visual processes in man; in DESMEDT Visual evoked potentials in man: new developments, pp. 16–89 (Oxford University Press, Oxford 1977).

SPERRY, R.W.: Lateral specialization of cerebral function in the surgically separated hemispheres; in MCGUIGAN and SCHOONOVER The psychophysiology of thinking, pp. 209–229 (Academic Press, New York 1973).

SPERRY, R.W.: Lateral specialization in the surgically separated hemispheres; in SCHMITT and WORDEN The neurosciences third study program, pp. 5–19 (MIT Press, Cambridge 1974).

SPERRY, R.W.; GAZZANIGA, M.S., and BOGEN, J.E.: Interhemispheric relationships: the neocortical commissures; syndromes of hemisphere disconnection; in VINKEN and BRUYN Handbook of clinical neurology, vol. 4, pp. 273–290 (North-Holland, New York 1969).

SPILKER, B.; KAMIYA, J.; CALLAWAY, E., and YEAGER, C.L.: Visual evoked responses in subjects trained to control alpha rhythms. Psychophysiology *5:* 683–695 (1969).

SQUIRES, K.C. and DONCHIN, E.: Beyond averaging: the use of discriminant functions to recognize event related potentials elicited by single auditory stimuli. Electroenceph. clin. Neurophysiol. *41:* 449–459 (1976).

SQUIRES, K.C.; HILLYARD, S.A., and LINDSAY, P.H.: Vertex potentials evoked during auditory signal detection: relation to decision criteria. Percept. Psychophys. *14:* 265–272 (1973).

SQUIRES, K.C.; WICKENS, C.; SQUIRES, N.C., and DONCHIN, E.: The effect of stimulus sequence on the waveform of the cortical event-related potential. Science *193:* 1142–1146 (1976).

SQUIRES, N.K.; DONCHIN, E.; SQUIRES, K.C., and GROSSBERG, S.: Bisensory stimulation: inferring decision-related processes from the P300 component. J. exp. Psychol. *3:* 299–315 (1977).

STACHOWIAK, F.J.; HUBER, W.; KERSCHENSTEINER, M.; POECK, K. und WENIGER, D.: Die globale Aphasie. Klinisches Bild und Überlegungen zur neurolinguistischen Struktur. J. Neurol. *214:* 75–87 (1977).

STANNERS, R.F.; HEADLEY, D.B., and CLARK, W.R.: The pupillary response to sentences: influences of listening set and deep structure. J. verbal Learn. Behav. *11:* 257–263 (1972).

STARR, A.: Clinical relevance of brain stem auditory evoked potentials in brain stem disorders in man; in DESMEDT Auditory evoked potentials in man. Psychopharmacology correlates of EPs, pp. 45–57 (Karger, Basel 1977).

STARR, A. and ACHOR, L.J.: Auditory brain stem responses in neurological disease. Archs Neurol., Chicago 32: 761–768 (1975).

STERNBERG, S.: The discovery of processing stages: extension of Donders' method; in KOSTER Attention and Performance, vol. 2 (North-Holland, Amsterdam 1969).

STOHR, P.E. and GOLDRING, S.: Origin of somatosensory evoked scalp responses in man. J. Neurosurg. 31: 117–127 (1969).

STUDDERT-KENNEDY, M.: Speech perception; in LASS Contemporary issues in experimental phonetics (Thomas, Springfield 1975).

STUDDERT-KENNEDY, M. and SHANKWEILER, D.: Hemispheric specialization for speech perception. J. acoust. Soc. Am. 48: 579–594 (1970).

SUBIRANA, A.: Handedness and cerebral dominance; in VINKEN and BRUYN Handbook of clinical neurology, vol. 4 (North-Holland, Amsterdam 1969).

SUGERMAN, A.A.; GOLDSTEIN, L.; MARJERISON, G., and STOLTZFUS, N.: Recent research in EEG amplitude analysis. Dis. nerv. Syst. 34: 162–166 (1973).

SURWILLO, W.W.: Interval histogram analysis of period of the electroencephalogram in relation to age during growth and development in normal children. Psychophysiology 12: 506–512 (1975).

SURWILLO, W.W.: Analysis of interval histograms of period of the electroencephalogram from homologous left and right derivations in verbal and nonverbal tasks. Physiol. Psychol. 4: 307–310 (1976).

SUTTON, S.: Specification of psychological variables in an averaged evoked potential experiment; in DONCHIN and LINDSLEY Average evoked potentials. NASA SP-191, pp. 237–297 (US Government Printing Office, Washington 1969).

SUTTON, S.; BRAREN, M., and ZUBIN, J.: Evoked potential correlates of stimulus uncertainty. Science 150: 1187–1188 (1965).

SUTTON, S.; TUETING, P.; ZUBIN, J., and JOHN, E.R.: Information delivery and the sensory evoked potential. Science 155: 1436–1439 (1967).

SUZUKI, T. and TAGUCHI, K.: Cerebral evoked response to auditory stimuli in young children during sleep. Ann. Otol. Rhinol. Lar. 77: 102–110 (1968).

SYNDULKO, K.: Relationships between motor potentials and CNV. Electroenceph. clin. Neurophysiol. 27: 706 (1969).

SYNDULKO, K.: Cortical slow potentials in humans during sensory and motor tasks; unpubl. diss., Los Angeles (1972).

SYNDULKO, K. and LINDSLEY, D.B.: Motor and sensory determinants of cortical slow potential shifts in man; in DESMEDT Attention, voluntary contraction and event-related cerebral potentials, pp. 97–131 (Karger, Basel 1977).

SZIRTES, J. and VAUGHAN, J.R.: Topographic analysis of speech-related cerebral potentials. Electroenceph. clin. Neurophysiol. 34: 754 (1973).

TAUB, J.M.; TANGUAY, P.E.; DOUBLEDAY, C.N., and CLARKSON, D.: Hemisphere and ear asymmetry in the auditory evoked response to musical chord stimuli. Physiol. Psychol. 4: 11–17 (1976).

TECCE, J.J.: Contingent negative variation and psychological processes in man. Psychol. Bull. 77: 73–108 (1972).

Tecce, J.J. and Scheff, N.M.: Attention reduction and suppressed direct-current potentials in the human brain. Science *164:* 331–333 (1969).

Teuber, H.L.: Effects of brain wounds implicating the right or left hemisphere in man; in Mountcastle Interhemispheric relations and cerebral dominance, pp. 131–157 (Johns Hopkins Press, Baltimore 1962).

Teuber, H.L.: Space perception and its disturbances after brain injury in man. Neuro-psychologia *1:* 47–57 (1963).

Teuber, H.L.: Why two brains? in Schmitt and Worden The neurosciences third study program, pp. 71–74 (MIT Press, Cambridge 1974).

Teuber, H.L. and Morrell, L.K.: Unpublished observations. Trans. Am. Neurol. Ass. *98:* 133 (1973).

Teuber, H.L. and Weinstein, S.: Performance on a form board task after penetrating brain injury. J. Psychol. *38:* 177–190 (1954).

Teyler, T.J.; Roemer, R.A.; Harrison, T.F., and Thompson, R.F.: Human scalp-recorded evoked-potential correlates of linguistic stimuli. Bull. Psychon. Soc. *1:* 333–334 (1973).

Thatcher, R.W. and John, E.R.: Information and mathematical quantification of brain states; in Burch and Altschuler Behavior and brain electrical activity, pp. 303–324 (Plenum, New York 1975).

Thieos, J.: Reaction time measurements in the study of memory processes: theory and data; in Bower The psychology of learning and motivation. Adv. Res. Theory, vol. 7 (Academic Press, New York 1974).

Thompson, R.F. and Bettinger, L.A.: Neural substrates of attention: contemporary theory and analysis (Appleton Century Crofts, New York 1970).

Thorne, J.; Bratley, P., and Dewar, H.: The syntactic analysis of English by machine; in Mitchie Machine intelligence, vol. 3 (Am. Elsevier, New York 1968).

Travis, L.E. and Knott, J.R.: Bilaterally recorded brain potentials from normal speakers and stutterers. J. Speech Dis. *2:* 239–241 (1937).

Treisman, A.: Strategies and models of selective attention. Psychol. Rev. *76:* 282–299 (1969).

Treub, S.E.: Infants sensitivity to vowel and tonal contrasts. Devl. Psychol. *9:* 91–96 (1973).

Trevarthen, C.: The psychobiology of speech development. Neurosci. Res. Prog. Bull. *12:* 570–585 (1974).

Tueting, P.; Sutton, S., and Zubin, J.: Quantitative evoked potential correlates of the probability of events. Psychophysiology *7:* 385–394 (1971).

Vaughan, H.G.: The analysis of scalp recorded brain potentials; in Thompson and Patterson Bioelectric recording techniques. Part B. Electroencephalography and human brain potentials, pp. 157–207 (Academic Press, New York 1974).

Vaughan, H.G.; Costa, L.D., and Ritter, W.: Topography of the human motor potential. Electroenceph. clin. Neurophysiol. *25:* 1–10 (1968).

Vaughan, H.G.; Katzman, R., and Taylor, J.: Alterations of visual evoked response in the presence of homonymous visual defects. Electroenceph. clin. Neurophysiol. *15:* 737–746 (1963).

Vaughan, H.G. and Ritter, W.: The sources of auditory evoked responses recorded from the human scalp. Electroenceph. clin. Neurophysiol. *28:* 360–367 (1970).

VELLA, E.J.; BUTLER, S.R., and GLASS, A.: Electrical correlate of right hemisphere function. Nature new Biol. *236:* 125–126 (1972).

WADA, J.A.: A new method for the determination of the side of cerebral speech dominance. A preliminary report on the intracarotid injection of sodium amytal in man. Med. Biol. *14:* 221–222 (1949).

WADA, J.A.; CLARKE, R., and HAMM, A.: Cerebral hemispheric asymmetry in humans. Archs Neurol., Chicago *32:* 239–246 (1975).

WADA, J.A. and DAVIS, A.E.: Fundamental nature of human infants brain asymmetries. Can. J. neurol. Sci. *4:* 203–207 (1977).

WADA, J.A. and RASMUSSEN, T.: Intracarotid injection of sodium amytal for the lateralization of cerebral speech dominance: experimental and clinical observations. J. Neurosurg. *17:* 266–282 (1960).

WALTER, V.J. and WALTER, W.G.: The central effects of rhythmic sensory stimulation. Electroenceph. clin. Neurophysiol. *1:* 57–86 (1949).

WALTER, W.G.: The location of cerebral tumors by electroencephalography. Lancet *ii:* 305–312 (1936).

WALTER, W.G.: The living brain (Duckworth, London 1954).

WALTER, W.G.; COOPER, R.; ALDRIDGE, V.J.; McCALLUM, W.C., and WINTER, A.L.: Contingent negative variation: an electric sign of sensorimotor association and expectancy in the human brain. Nature, Lond. *203:* 380–384 (1964).

WANNER, E.: On remembering, forgetting and understanding sentences: a study of the deep structure hypothesis (Mouton, The Hague 1974).

WANNER, E. and MARATSOS, M.: An augmented transition network model of relative clause comprehension (unpubl. paper, 1974).

WARREN, L.R.; PELTZ, L., and HAUETER, E.S.: Patterns of EEG alpha during word processing and relations to recall. Brain Lang. *3:* 283–291 (1976).

WARREN, R.M.: Verbal transformation effect and auditory perceptual mechanisms. Psychol. Bull. *70:* 261–270 (1968).

WARRINGTON, E.K. and PRATT, R.T.C.: Language laterality in left-handers assessed by unilateral E.C.T. Neuropsychologia *11:* 423–428 (1973).

WEBSTER, W.: Hemispheric asymmetry in cats; in HARNAD, DOTY, GOLDSTEIN, JAYNES and KRAUTHAMER Lateralization in the nervous system, pp. 471–480 (Academic Press, New York 1977).

WEERTS, T.C. and LANG, P.J.: The effects of eye fixation and stimulus and response location on the contingent negative variation (CNV). Biol. Psychol. *1:* 1–19 (1973).

WEINBERG, H. and PAPAKOSTOPOULOS, D.: The frontal CNV; its dissimilarity to CNVs recorded from other sites. Electroenceph. clin. Neurophysiol. *39:* 21–28 (1975).

WEINGARTEN, K.: Tics; in VINKEN and BRUYN Handbook of clinical neurology, vol. 6, pp. 782–808 (North Holland, Amsterdam 1968).

WERNICKE, K.: A psychological study on an anatomical basis; in COHEN and WARTOFSKY Boston studies in the philosophy of science, vol. 4 (Reidel, Dordrecht 1969).

WHITE, M.J.: Laterality differences in perception: a review. Psychol. Bull. *72:* 387–405 (1969).

WILKE, J.T. and LANSING, R.W.: Variations in the motor potential with force exerted during voluntary arm movements in man. Electroenceph. clin. Neurophysiol. *35:* 259–265 (1973).

Winer, B.J.: Statistical principles in experimental design (McGraw-Hill, New York 1962).

Witelson, S.F.: Hemispheric specialization for linguistic and nonlinguistic tactual perception using a dichotomous stimulation technique. Cortex 10: 3–17 (1974).

Witelson, S.F. and Pallie, W.: Left hemisphere specialization for language in the newborn: anatomical evidence for asymmetry. Brain 96: 641–646 (1973).

Wood, C.C.: Parallel processing of auditory and phonetic information in speech perception. Percept. Psychophys. 15: 501–508 (1974).

Wood, C.C.: Auditory and phonetic levels of processing in speech perception: neurophysiological and information-processing analyses. J. exp. Psychol. 104: 3–20 (1975a).

Wood, C.C.: A normative model for redundancy gains in speech discrimination; in Restle, Shiffrin, Castellan, Lindman and Pisoni Cognitive theory, vol. 1 (Erlbaum, Potomac 1975b).

Wood, C.C.; Goff, W.R., and Day, R.S.: Auditory evoked potentials during speech perception. Science 173: 1248–1251 (1971).

Woods, W.: Transition network grammars for natural language analysis. Commun. ACM 13: 591–606 (1970).

Wright, P. and Kahneman, D.: Evidence for alternative strategies in sentence retention. Q. Jl exp. Psychol. 23: 197–213 (1971).

Yagi, A.; Bali, L., and Callaway, E.: Optimum parameters for the measurement of cortical coupling. Physiol. Psychol. 4: 33–38 (1976).

Yingling, C.D. and Skinner, J.E.: Gating of thalamic input to cerebral cortex by nucleus reticularis thalami; in Desmedt Attention, voluntary contraction and event-related cerebral potentials, pp. 70–96 (Karger, Basel 1977).

Zaidel, E.: Auditory vocabulary of the right hemisphere following brain bisection or hemidecortication. Cortex 12: 191–211 (1976).

Zangwill, O.L.: Cerebral dominance and its relation to psychological function (Oliver & Boyd, Edinburgh 1960).

Zimmermann, G.N. and Knott, J.R.: Slow potentials of the brain related to speech processing in normal speakers and stutterers. Electroenceph. clin. Neurophysiol. 37: 599–607 (1974).

Subject Index

Active touch 180
Adaptation level 45
Adiadochokinesia 32
Affective states 170
Agrammatism 29, 33
Akinesia 29, 33
Alpha asymmetry 141, 155, 217
- distribution 148, 217
- phase difference 217
Ambiguous linguistic stimuli 3, 49
Anarthria 8
Animal communication 17
Ansa lenticularis 34
Anxiety 167
Aphasia 1, 19, 28
Apomorphin 33
Arousal 46, 166, 170, 252
Articulation 7
Association cortex 19
Asymmetry in event-related potentials 9, 23, 38, 48, 58, 89, 103, 132, 147, 151, 179, 184, 192, 223, 234
Athetosis 33
Attention 3, 173, 208, 243
Auditory event-related potentials 25, 128, 158, 185, 191, 206, 214, 223, 233
- -phonetic levels of processing 73
Augmented transition network 13
Autospectrum 127
Averaging 25, 38, 79, 151, 173

Background EEG 63, 70, 79, 162, 212
Ballistic movements 33
Bandpass of recording system 40, 48, 75, 88, 106, 114, 173, 190, 207, 244
Basal ganglia asymmetry 28, 33
Behaviorism 2
Bereitschaftspotential 56, 89, 103, 225
Biofeedback 219
Blindness 168
Block design tasks 143, 153, 219
Braille reading 168
Brain lesions 23, 31, 60, 140, 173, 183, 188, 213
- stimulation 188
Brainstem nuclei 29
Broca area 1, 19, 28, 92, 106, 113

Cannabis 167
Carotid amytal test 105, 128, 134, 211
Categorical perception 5
Central-moments technique 220
Cerebellum 28, 32
Children ERP 191
Chimpanzee language 18
Choice reaction time 210, 221
Choreiform movements 32
CNV 81, 108, 112, 124, 172, 205, 225, 252
Cognitive ERP correlates 58, 65, 71, 110, 184, 206

Subject Index